THERMOELECTRIC POWER OF METALS

THERMOELECTRIC POWER OF METALS

Frank J. Blatt, Peter A. Schroeder, and Carl L. Foiles

Michigan State University
East Lansing, Michigan

and

Denis Greig

University of Leeds
Leeds, England

PLENUM PRESS · NEW YORK AND LONDON

Library of Congress Cataloging in Publication Data
Main entry under title:

Thermoelectric power of metals.

Includes bibliographical references and index.
1. Thermoelectricity. 2. Metals—Thermal properties. 3. Metals—Electric proper-
ties. I. Blatt, Frank J.
QC621.T47   537.6'5   76-20706
ISBN-13:978-1-4613-4270-0   e-ISBN-13:978-1-4613-4268-7
DOI: 10.1007/978-1-4613-4268-7

© 1976 Plenum Press, New York
Softcover reprint of the hardcover 1st edition 1976

A Division of Plenum Publishing Corporation
227 West 17th Street, New York, N.Y. 10011

| Preface

Thermoelectric and related transport properties of metals have been a source of information and, also, exasperation to physicists for over a century. Perhaps the principal reasons for interest in these phenomena are their sensitivity to composition, structure and external fields and, until fairly recently, the distressing fact that often even gross experimental features such as the sign of the thermopower eluded theoretical understanding.

During the past two decades many of the previously perplexing aspects of thermoelectricity have yielded to more sophisticated theoretical treatment. As a result of this effort and concomitant experimental work using advanced measurement techniques, there is now good reason to believe that thermoelectric phenomena can shed much light on the interactions between electrons and phonons, impurities, and other defects.

The last few years have witnessed new and fascinating developments that promise to stimulate new activity in this field. In contrast to the more conventional transport properties, second- and high-order contributions in electron scattering theory appear to play a profound role in thermoelectricity—the controversy surrounding ordinary and "phony" phonon drag is far from resolved; the startlingly large effect of magnetic fields on the thermopower of metals appears to be linked intimately to scattering anisotropy; quantum oscillations of thermopower are orders of magnitude larger than corresponding oscillations of the magnetoresistance; a new approach to thermoelectric studies allows extension of thermopower measurements into the millikelvin region of temperature; finally, the advent of superconducting detection devices permits the precise measurement of extremely small voltages, an essential requirement in this field.

Since several review articles and a book on thermoelectricity have appeared in the last few years, we have placed special emphasis on those facets of the field that appear to be of current interest and promise to hold the attention of experimenters and theorists for at least a few years. Following a brief introductory chapter and a cursory survey of the electronic theory of metallic conduction is a chapter on experimental techniques which stresses Josephson junction devices as used in thermopower studies. This is followed by a detailed discussion of phonon drag. For comparison purposes, the provocative theoretical work of Nielson and Taylor on the effect of second-order scattering on diffusion thermopower is included in the same chapter. The next two chapters are closely related; one is concerned with the thermopower of transition metals, with special attention to the relationship between this and their magnetic properties, and the other on dilute magnetic alloys wherein the still not fully resolved Kondo effect is highlighted. The final chapter is devoted to effects of pressure and of magnetic fields on thermoelectricity, areas where there have been a number of fascinating developments in the last years.

Although we have not compiled a complete bibliography, we have made an attempt to include a large and representative list of references. We apologize, in advance, for the many omissions.

Most of the illustrations in this book are reproduced from articles that have appeared in the periodical literature or from other books; the source of each such illustration appears in the figure caption. We are grateful to the many authors and publishers who gave their permission to reproduce these illustrations.

In the preparation of the manuscript we have had the benefit of discussions with many individuals. One deserves particular mention; Dr. Jon Opsal not only gave it a thorough critical reading, but he also made numerous valuable suggestions and is largely responsible for Section 4.7.

We also express our thanks to Mrs. Jean Strachan for her patience and skill in typing the manuscript.

June 1976   F.J.B.

East Lansing, Michigan   P.A.S.
                                                                            C.L.F.
Leeds, England   D.G.

Contents

| Notation

Symbol	Definition	First used on page	Equation
$\mathbf{a}_n$	Direct lattice vector	16	2.12
c	Mole fraction of point defect	114	4.41
C_e	Electronic specific heat	32	2.25a
C_g	Lattice specific heat	28	2.34
C_m	Magnon specific heat	171	5.23
e	Electronic charge	7	—
E	Thermal emf	94	4.20
$\mathbf{E}$	Electric field vector	6	1.10
E_L	Energy of virtual bound state	194	—
f	Non equilibrium distribution function for electrons	17	2.17
f_0	Equilibrium Fermi–Dirac distribution	15	2.5
f_1	$f - f_0$	18	2.18
$\mathbf{F}$	External driving force	17	—
G	Thermoelectric ratio	49	3.1
H	Magnetic field	7	1.11
j	Polarization index	14	—
J	Exchange integral	141	—
$\mathbf{J}$	Electrical current density	3	1.6
J_i^μ	Generalized flux of type i	6	1.11
k	Boltzmann constant	15	—
$\mathbf{k}$	Electron wave vector	14	—

Symbol	Definition	First used on page	Equation
$\mathbf{k}_f$	Fermi wave vector	124	—
$\boldsymbol{K}_n, K_n$	Transport integrals	19	2.23
$\mathbf{K}_n$	Reciprocal lattice vector	16	2.12
$l(\varepsilon)$,	Electron mean free path	23	2.29
$\mathbf{l}(\mathbf{k})$	Electron mean free path	25	2.30
l_g	Phonon mean free path	28	2.34
L	Wiedemann–Franz ratio	112	4.38
L_0	Lorenz number	11	—
$\mathscr{L}_{ij}^{\mu\nu}$	Transport tensor	6	1.11
m	Electron mass	14	2.1
M	Atomic mass	114	4.42
$M_s(T)$	Saturation magnetization	162	—
n_0	Electron density	15	2.6
$N(\varepsilon)$	Density of states for electrons	15	2.4
$N_d(\varepsilon)$	Density of states for d electrons	136	—
$N_s(\varepsilon)$	Density of states for s electrons	146	—
$N_0(\mathbf{q}), N(\mathbf{q})$	Equilibrium and perturbed phonon distributions	88	4.4, 4.3
$\mathbf{p}$	Electron momentum	15	—
$\mathbf{q}$	Phonon wave vector	14	—
$\mathbf{q}_{min}$	Minimum $\mathbf{q}$ for Umklapp process	35	—
$\mathbf{Q}$	Heat current density	6	1.10
$\dot{Q}_p$	Peltier heat	3	1.6
$\mathbf{r}$	Position vector	17	—
$\mathbf{S}$	Absolute thermopower tensor	2	1.2
S_A, S_B	Absolute thermopowers for isotropic material	2	1.4
S_{AB}	Thermopower of thermocouple	1	1.1
S_d	Diffusion thermopower	19	2.25
S_d^d	of d electrons	146	5.7
S_d^{ee}	for electron–electron scattering	152	—
S_d^i	for phonon scattering	38	2.43
S_d^j	for jth scattering mechanism	38	2.43
S_d^s	of s electrons	146	5.7
S_d^r, S_{d0}	for residual impurity scattering	99, 158	5.12
S_g	Phonon drag thermopower	32	2.35

Symbol	Definition	First used on page	Equation
S_g^0	for electron phonon scattering only	33	2.37
S_g^N	for normal scattering processes	35	—
S_g^U	for Umklapp scattering processes	35	—
S_{g_i}	for ith region of Fermi surface	108	4.36
S_m	Magnon drag thermopower	170	5.23
S_0	$\pi^2 k^2 T_B / 3e\eta$	123	4.47
T	Temperature	1	1.1
T_C	Curie temperature	160	—
T_K	Kondo temperature	205	—
T_N	Néel temperature	146	—
v	Velocity of sound	93	4.17
$\mathbf{v}$	Particle velocity	7	—
V	Potential	1	1.1
V	Volume	14	2.2
$V(j\mathbf{q})$	Velocity associated with phonon $(j\mathbf{q})$	88	—
W	Electronic thermal resistivity	39	2.46
W_{ee}	for electron–electron scattering	152	—
W_i	for electron–phonon scattering	39	2.46
W_j	for scattering from impurity j	39	2.46
W_r	for residual impurity scattering	99	2.46
x	S/S_0	123	4.47
$x_1, \Delta x_1$	$x = x_1 + \Delta x_1$	123	4.48
x_i, x_j	Friedel parameters	38	2.42
x	$\hbar\omega/kT$	96	4.21
Z_i	Valence of host	40	—
Z_j	Valence of impurity	40	—
Z	$Z_s - Z_i$	150	—
ΔZ	Charge on ion—charge on host ion	194	—
$\alpha(j\mathbf{q}; \mathbf{k}l, \mathbf{k}'l')$	Phonon scattering probability	89	4.8
α_j	$d \ln \rho_j / d \ln V$	45	2.47
γ	Grüneisen constant	46	2.49
γ	$(\eta - \varepsilon_d)/kT$	147	—

Symbol	Definition	First used on page	Equation
Γ	Half width of virtual bound state	195	—
δ_l	Phase shift of lth partial wave	195	—
Δ	$\Gamma/2$	203	—
Δ_0	Atomic volume	114	4.41
ε	Electron energy	7	—
$\mathscr{S}$	Surface area in k space	17	2.16
η	Fermi energy	7	—
θ_D	Debye temperature	23	—
θ_L	Debye temperature for longitudinal waves	97	4.27
θ_T	Debye temperature for transverse waves	97	4.27
θ^*	$\hbar v \mathbf{q}_{\min}/k$	93	4.18
κ	Electronic thermal conductivity ($j=0$)	19	2.24
κ_e	Electronic thermal conductivity ($\mathbf{E}=0$)	8	—
κ_g	Lattice thermal conductivity	28	2.34
Λ_i	Phonon mean free path for impurity scattering	96	4.22
Λ_e	Phonon mean free path for electron scattering	96	4.22
μ	Thomson coefficient	3	1.7
μ_e	Effective magnetic moment	204	6.10
Π_A, Π_{AB}	Peltier coefficients	3	1.6
Π_e	$(=G^{-1})$	8	—
Π_t	$(=\Pi_A)$	8	—
ρ	Resistivity	3	1.7
ρ_{ee}	electron–electron scattering	138	—
ρ_i	phonon scattering	37	2.39
ρ_j	scattering by impurity j	37	2.39
ρ_r	residual impurity scattering	196	—
σ	Electrical conductivity	11	—
σ_a	adiabatic	11	—
σ_d	of d electrons	137	5.1
σ_i	of ith region of Fermi surface	108	4.35

Symbol	Definition	First used on page	Equation
σ_s	of s electrons	137	5.1
σ_t	Isothermal electrical conductivity	19	2.24a
τ	Electron relaxation time	18	2.18
τ_{dd}	for d–d scattering	146	—
τ_{sd}	for s–d scattering	146	—
$\tau_0(\uparrow),\ \tau_0(\downarrow)$	for impurity scattering of electrons with spins ($\uparrow$) and ($\downarrow$)	154	—
τ_g	Mean phonon relaxation time	90	4.12
τ_p	Phonon relaxation time		—
τ_{pe}	for electron scattering	33	2.37
τ_{pi}	for impurity scattering	113	4.40
τ_{pp}	for phonon scattering	115	4.43
τ_p'	for all scattering processes other than phonon–electron	33	2.37
ϕ	Magnetic flux	74	—
ϕ	Scattering temperature	97	4.25
ϕ_0	Flux quantum	74	—
$\mathfrak{S}$	Entropy	7	1.12

1 | Introduction

1.1 Seebeck, Peltier, and Thomson Effects

The phenomenon of thermoelectricity was first observed in 1826 by Seebeck ['26S1], who found that a current will flow in a closed circuit made of two dissimilar metals when the two junctions are maintained at different temperatures. Today, when we speak of the Seebeck effect, we generally envisage an open circuit, such as that shown in Fig. 1.1. The voltage $\Delta V = V_b - V_a$ is the thermoelectric voltage developed by this couple, and the thermoelectric power of the couple is defined by

$$S_{AB} = \lim_{\Delta T \to 0} (\Delta V / \Delta T) \qquad (1.1)$$

Since the signs of thermoelectric voltages and of the thermoelectric power are an endless source of confusion, we first derive in detail the relationships for the thermocouple of Fig. 1.1. The terminals a and b are assumed to be at the same temperature, and junctions c and d at temperatures T_2 and T_1 as shown. Although the thermoelectric power of a couple evidently involves the difference in the response of two dissimilar metals to a temperature gradient, it is nevertheless possible, and also convenient, to define the absolute thermoelectric power S, which is a unique physical

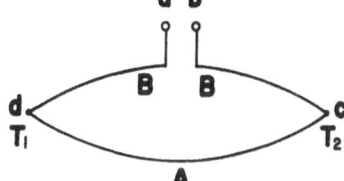

Fig. 1.1. An open circuit that displays the Seebeck effect.

1

property of a particular material, by the relation

$$\mathbf{E} = \mathbf{S}\nabla T \tag{1.2}$$

where $\mathbf{E}$ is the electric field in the material and ∇T the temperature gradient. Since $\mathbf{E}$ and ∇T are vector quantities, $\mathbf{S}$ is generally a tensor. For the moment, however, we shall disregard this complication and assume that the conductors A and B of Fig. 1.1 have cubic symmetry.

From Eq. (1.2) we have

$$-\nabla V = \mathbf{S}\nabla T \quad \text{or} \quad dV = -S\,dT \tag{1.3}$$

Thus,

$$\begin{aligned}
\Delta V = V_b - V_a &= (V_b - V_c) + (V_c - V_d) + (V_d - V_a) \\
&= \int_c^b dV + \int_d^c dV + \int_a^d dV \\
&= -\int_c^b S_B\,dT - \int_d^c S_A\,dT - \int_a^d S_B\,dT \\
&= -\int_c^d S_B\,dT - \int_d^c S_A\,dT \\
&= \int_{T_1}^{T_2} (S_B - S_A)\,dT
\end{aligned} \tag{1.4}$$

If, then, the absolute thermoelectric power, hereafter abbreviated TEP, of metal A is zero, the terminal of the circuit connected to the high-temperature junction will be at a positive potential with respect to the low-temperature terminal if S_B is positive. Evidently the thermoelectric power of the couple, defined by Eq. (1.1), is given by

$$S_{AB} = S_B - S_A \tag{1.5}$$

The second of the three common thermoelectric effects, named after its discoverer Jean Charles Athanase Peltier ['34P1], relates to the heat reversibly liberated (or absorbed) at a junction between two dissimilar metals when a current passes through the junction. This heat is reversible in the sense that if Peltier heat is evolved when current passes from conductor A to conductor B, the same amount of heat will be absorbed if the direction of current flow is reversed, and is thus easily distinguished experimentally from the Joule heat, which is independent of the direction of current flow. As with the TEP, one can define an absolute Peltier coefficient even though the observation of the effect of necessity involves two different conductors. In terms of the absolute Peltier coefficients Π_A and Π_B, one has

$$\dot{Q}_P = \Pi_{AB}J \tag{1.6}$$

where $\Pi_{AB} = \Pi_B - \Pi_A$; $\dot{Q}_P$ is the reversible heat absorbed per unit time and unit cross-sectional area when a current of density J passes from conductor A to conductor B.

As MacDonald [62M1] strongly emphasized in his book, the Peltier effect is not a contact phenomenon in the usual sense. The reversible heat liberated or absorbed depends not on the nature of the contact but on the intrinsic properties of the two conductors, namely, their absolute Peltier coefficients.

Since both the Seebeck and Peltier effects can only be observed when two dissimilar conductors are employed, one may well wonder how the absolute coefficients can ever be ascertained. The answer is provided by the third thermoelectric phenomenon, the Thomson effect, which is the reversible evolution (or absorption) of heat in a homogeneous conductor that carries an electric current and in which a temperature gradient is also maintained. The rate of heat production per unit volume is then given by

$$\dot{Q} = \rho J^2 - \mu J \nabla T \tag{1.7}$$

Here ρ is the resistivity, J the current density, and μ the Thomson coefficient. The first term in Eq. (1.7) is the Joule heat; the second term is the reversible Thomson heat. The Thomson coefficient μ is positive if heat is absorbed from the surroundings when J and ∇T are parallel.

In contrast to the Seebeck and Peltier effects, the Thomson effect involves only a single homogeneous conductor. Hence, μ can be determined independently for any conductor, and since μ, S, and Π are related to each other by the Kelvin relations,

$$\mu = T\frac{dS}{dT} \tag{1.8a}$$

$$\Pi = TS \tag{1.8b}$$

the absolute TEP and Peltier coefficients can be calculated if the Thomson coefficient is known as a function of temperature.

Equation (1.8a) follows from energy conservation. The second of the Kelvin relations was originally derived on the basis of thermodynamic arguments ['82K1, 34B1]. Since in a normal metal current flow is necessarily an irreversible phenomenon, Eq. (1.8b) was suspect for some time until its validity was established using the more sophisticated techniques of irreversible thermodynamics. Equation (1.8b) is, in fact, a direct consequence of the Onsager reciprocity relations [31O1, 48C1, 52C1, 53D1]. For the moment it will suffice to say that Eqs. (1.8a) and (1.8b) are theoretically sound and have been subjected to experimental verification.

In Chapter 3 we shall discuss in detail various techniques for the measurement of thermoelectric phenomena. Here we shall only note that the TEP of a substance is obtained from the Thomson coefficient by integration, i.e.,

$$S(T) = \int_0^T \frac{\mu(T')}{T'} \, dT' \tag{1.9}$$

Of course, the accurate measurement of the Thomson coefficient is no simple task, especially at very low temperatures. Fortunately, one of the many remarkable properties of superconductors may be used here to advantage. It can be shown from general theoretical arguments that the TEP of a superconductor is identically zero. Hence, the TEP of a normal metal at low temperatures may be obtained by forming a couple using the normal metal and a superconductor and measuring the thermopower of this couple, which is then due to the normal arm of the couple only. The TEP of lead

Table 1.1. The Absolute Thermoelectric Power of Pure Lead between 0 and 300 K

T, °K	S, μV/deg	T, °K	S, μV/deg
0	0	60	-0.77_9
5	0	70	-0.78_4
7.5	-0.22_1	80	-0.79_4
8	-0.25_7	90	-0.82_4
8.5	-0.29_7	100	-0.86_5
9	-0.34_3	113.2	-0.91
10	-0.43_4	133.2	-0.96
11	-0.51_6	153.2	-1.02
12	-0.59_3	173.2	-1.06
14	-0.70_6	193.2	-1.10_5
16	-0.77_1	213.2	-1.15
18	-0.78_{45}	233.2	-1.18
20	-0.78_4	253.2	-1.21
30	-0.77_4	273.2	-1.25
40	-0.76_4	293.2	-1.27_5
50	-0.77_4		

Table 1.2. The Absolute Thermoelectric Powers for Some Metals [58C2]

				Metal			
T, °K	Cu	Ag	Au	Pt	Pd	W	Mo
100	1.19	0.73	0.82	4.29	2.00	–	–
150	1.12	0.85	1.02	1.32	−1.63	–	–
200	1.29	1.05	1.34	−1.27	−4.85	–	–
273	1.70	1.38	1.79	−4.45	−9.00	0.13	4.71
300	1.83	1.51	1.94	−5.28	−9.99	1.07	5.57
400	2.34	2.08	2.46	−7.83	−13.00	4.44	8.52
500	2.83	2.82	2.86	−9.89	−16.03	7.53	11.12
600	3.33	3.72	3.18	−11.66	−19.06	10.29	13.27
700	3.83	4.72	3.43	−13.31	−22.09	12.66	14.94
800	4.34	5.77	3.63	−14.88	−25.12	14.65	16.13
900	4.85	6.85	3.77	−16.39	−28.15	16.28	16.86
1000	5.36	7.95	3.85	−17.86	−31.18	17.57	17.16
1100	5.88	9.06	3.88	−19.29	−34.21	18.53	17.08
1200	6.40	10.15	3.86	−20.69	−37.24	19.18	16.65
1300	6.91	–	3.78	−22.06	−40.27	19.53	15.92
1400	–	–	–	−23.41	−43.30	19.60	14.94
1600	–	–	–	−26.06	−49.36	18.97	12.42
1800	–	–	–	−28.66	−55.42	17.41	9.52
2000	–	–	–	−31.23	−61.48	15.05	6.67
2200	–	–	–	–	–	12.01	4.30
2400	–	–	–	–	–	8.39	2.87

between 7 and 18 K was determined in this manner by Christian *et al.* [58C1]. Two decades previously Borelius [28B1, 30B1, 31B1, 32B1] had measured the Thomson coefficient of lead from very low temperatures to about 300 K.* The TEP of lead, the currently accepted standard for thermoelectric measurements, is based on this work and is shown in Table 1.1. At higher temperatures the noble metals Cu, Ag, or Au can serve as "standards" and are to be preferred over the transition metals Pt or W. Their thermopowers are substantially larger so that a given percentage error corresponds to a larger absolute errror in Pt than in Au. Unfortunately, comparison of values for S of platinum quoted by various sources [47N1, 48L1, 58C2, 66H2] as well as for Au [60P1, 64H1, 58C2, 47N1] reveals discrepancies of about 10%.

Table 1.2 is a reproduction of the tabulation presented by Cusack and Kendall [58C2]. It is important to point out that since these values are based

* A detailed description of the experiments of Borelius is given in Chapter 3.

on the lead data and the latter, in turn, were obtained at a time when measurement techniques were much less precise than they are today, it is fair to say that a new determination of the TEP of lead is long overdue. It is, therefore, gratifying to know that a new determination of the Thomson coefficient of lead is currently in progress.* Thus, roughly a half-century later the work of Borelius will be subjected to independent verification.

1.2   Transport Coefficients and Onsager Relations

In the preceding section we have treated the TEP as a scalar proportionality constant relating the thermoelectric field to the temperature gradient. As we mentioned earlier, S is in fact a second-rank tensor, but its tensor properties can become apparent only if we perform suitable measurements, for example, on single crystals of lower than cubic symmetry or in the presence of a transverse magnetic field. Such measurements, though more difficult than on polycrystalline wires, are also of greater fundamental interest and yield more detailed information. Let us then write the equations for charge and heat current:

$$\mathbf{J} = \mathscr{L}_{11}\mathbf{E} + \mathscr{L}_{12}\nabla T$$
$$\mathbf{Q} = \mathscr{L}_{21}\mathbf{E} + \mathscr{L}_{22}\nabla T \tag{1.10}$$

where $\mathbf{J}$ and $\mathbf{Q}$ are the electrical and heat current densities and $\mathscr{L}_{ij}$ are 3×3 tensors. Although the electric field $\mathbf{E}$ and temperature gradient ∇T appear at first sight to be reasonable choices for the driving forces, they are in fact not the best. The proper selection is dictated by the Onsager relations.

In general, we may think of the fluxes as the response of the system to generalized forces. If the response is linear, a basic assumption of transport theory, we can then write

$$J_i^\mu = \sum_{j\nu} L_{ij}^{\mu\nu}(H)F_j^\nu \tag{1.11}$$

Here J_i^μ denotes a flux of type i, the superscripts indicate the coordinate component, and we have specifically included the possible presence of a magnetic field H in writing the generalized transport coefficient $L_{ij}^{\mu\nu}$.

Forces and fluxes are canonically conjugate, in the sense of irreversible thermodynamics, if the rate of entropy production arising from the fluxes is

---

* R. Roberts (private communication).

given by

$$\mathfrak{S} = \sum_i \sum_\mu F_i^\mu J_i^\mu \tag{1.12}$$

If the forces and fluxes are canonically conjugate, the Onsager relation

$$L_{ij}^{\mu\nu}(H) = L_{ji}^{\nu\mu}(-H) \tag{1.13}$$

is valid.

There are two distinctly different kinds of forces which may induce a flux. The first, and more familiar, are electromechanical, for example, the force acting on a charge in an electric field, or gravitational or other mechanical forces acting on a mass. The second kind of force has its origin in statistics; it is not a "force" in the usual sense, but normally arises from concentration gradients and results in diffusive fluxes. It is convenient to relate both forces to potential gradients: the electrostatic force acting on a charge to the electrical potential gradient and the statistical force to the gradient of the chemical potential. Both forces must be included in Eq. (1.11). In all of our discussions we shall be concerned with an assembly of electrons, i.e., a Fermi system, for which the chemical potential is the Fermi energy η. It is then common practice to introduce the *electrochemical potential* $\eta + e\phi$, where ϕ is the electrostatic potential, and to define an *effective electric field* $\mathbf{E} = -\nabla(\phi + \eta/e)$, which is the field that is normally measured.*

To obtain the appropriate force conjugate to the electric current density **J**, we recall that the rate of irreversible heat production is given by $Q = \mathbf{J} \cdot \mathbf{E}$. Since $d\mathfrak{S} = TdQ$, the force in Eq. (1.12) must then be E/T. The chemical potential also enters into the heat flow, which is the flow of energy minus the flow of free energy. Thus, the heat flow for a collection of N Fermi particles is given by

$$\mathbf{Q} = \sum_{n=1}^N \mathbf{v}_n \varepsilon_n - \sum_{n=1}^N \mathbf{v}_n \eta = \sum_{n=1}^N \mathbf{v}_n (\varepsilon_n - \eta)$$

where $\mathbf{v}_n$ and ε_n are the velocity and energy of the nth particle. To obtain the force conjugate to **Q**, we associate an entropy flow $\mathbf{Q}/T$ with the heat flux. From the equation of continuity we then have

$$\mathfrak{S} = \nabla \cdot (\mathbf{Q}/T) = \mathbf{Q} \cdot \nabla(1/T) \tag{1.14}$$

from which we see that the correct conjugate force to **Q** is not ∇T but $\nabla(1/T) = -(1/T^2)\nabla T$.

---

* Hereafter the symbol **E** in Eq. (1.10) and elsewhere will always represent this effective field.

Table 1.3. The Twelve Experimental Transport Parameters for Cubic Materials[a]

Boundary condition	Symbol	Defining equation	Expressions in terms of K_n [see Eq. (2.23)]	Results in elastic scattering limit
Isothermal $\nabla T = 0$	Π_t^b	$\left(\dfrac{Q}{j}\right)_{\nabla T=0}$	$\dfrac{K_1}{eK_0}$	$eL_0T^2\dfrac{\sigma_0'}{\sigma_0}$
	σ_t	$\left(\dfrac{j}{E}\right)_{\nabla T=0}$	$e^2 K_0$	σ_0
	Ω_t	$\left(\dfrac{Q}{E}\right)_{\nabla T=0}$	eK_1	$eL_0T^2\sigma_0'$
Isoelectric $\mathbf{E} = 0$	Π_e^b	$\left(\dfrac{Q}{j}\right)_{\mathbf{E}=0}$	$\dfrac{K_2}{eK_1}$	$\dfrac{\sigma_0}{e\sigma_0'}$
	Σ_e	$-\left(\dfrac{j}{\nabla T}\right)_{\mathbf{E}=0}$	$\dfrac{eK_1}{T}$	$eL_0T\sigma_0'$
	κ_e	$-\left(\dfrac{Q}{\nabla T}\right)_{\mathbf{E}=0}$	$\dfrac{K_2}{T}$	$L_0T\sigma_0$

Static $j=0$			
S	$\left(\dfrac{E}{\nabla T}\right)_{j=0}$	$\dfrac{K_1}{eK_0T}$	$eL_0T\dfrac{\sigma_0'}{\sigma_0}$
Ω	$\left(\dfrac{Q}{E}\right)_{j=0}$	$eK_1 - \dfrac{eK_2K_0}{K_1}$	$-\dfrac{\sigma_0^2}{e\sigma_0'} + [eL_0T^2\sigma_0']^c$
κ	$-\left(\dfrac{Q}{\nabla T}\right)_{j=0}$	$\dfrac{K_2}{T} - \dfrac{K_1^2}{K_0T}$	$L_0T\sigma_0 - [(eL_0T^2\sigma_0')^2/\sigma_0T]^c$

Adiabatic $Q=0$			
S_a	$-\left(\dfrac{E}{\nabla T}\right)_{Q=0}$	$-\dfrac{K_2}{eK_1T}$	$-\dfrac{\sigma_0}{eT\sigma_0'}$
σ_a	$\left(\dfrac{j}{E}\right)_{Q=0}$	$e^2K_0 - \dfrac{e^2K_1^2}{K_2}$	$\sigma_0 - [e^2L_0T^2(\sigma_0')^2/\sigma_0]^c$
Σ_a	$-\left(\dfrac{j}{\nabla T}\right)_{Q=0}$	$\dfrac{eK_1}{T} - \dfrac{eK_0K_2}{K_1T}$	$-\dfrac{\sigma_0^2}{eT\sigma_0'} + [eL_0T\sigma_0']^c$

[a] The corresponding transport ratios in terms of the transport coefficients K_n [see Eqs. (2.22), (2.23), and (2.24)] are given. The last column gives the results in the limit of elastic scattering, when $\kappa_e = \kappa = \kappa_0$, $\sigma_t = \sigma_a = \sigma_0$, and $\kappa_0/\sigma_0T = L_0 = (\pi^2/3)(k/e)^2$; $\sigma_0' \equiv (\partial\sigma_0/\partial\varepsilon)_{\varepsilon=\eta}$.

[b] In this text we shall use the symbol Π to denote Π_t; we shall use the more common symbol $G \equiv 1/\Pi_e$ rather than Π_e hereafter.

[c] The terms in square brackets [ ] are corrections of order $(kT/\eta)^2$ and are, therefore, negligibly small in metals.

Thus, in order that we may apply the Onsager relation (1.13), we must rewrite the transport equations as follows:

$$\mathbf{J} = \mathbf{L}_{11} \frac{1}{T} \mathbf{E} - \mathbf{L}_{12} \frac{1}{T^2} \nabla T \tag{1.15}$$

$$\mathbf{Q} = \mathbf{L}_{21} \frac{1}{T} \mathbf{E} - L_{22} \frac{1}{T^2} \nabla T \tag{1.16}$$

It now remains to relate the above coefficients to the experimentally measured quantities, such as electrical conductivity, thermal conductivity, thermoelectric power, Hall coefficient, etc. To do this properly, it is important that the "boundary conditions" be clearly specified; for example, in principle the electrical conductivity can be measured while maintaining $\nabla T = 0$ or while maintaining $\mathbf{Q} = 0$. The first of these conductivities is the isothermal electrical conductivity σ_t; the second is the adiabatic electrical conductivity σ_a.

For the isothermal conductivity σ_t, we obtain directly from Eq. (1.15) and the definition $\mathbf{J} = \sigma \mathbf{E}$

$$\sigma_t = \frac{1}{T} \mathbf{L}_{11} \tag{1.17}$$

The thermoelectric power, determined under open-circuit conditions, i.e., with $\mathbf{J} = 0$, is related to $\mathbf{L}_{12}$. From Eqs. (1.15) and (1.16) we have

$$\mathbf{S} = \mathbf{L}_{12} \mathbf{L}_{11}^{-1} / T \tag{1.18}$$

Next, if we impose the isothermal condition, Eq. (1.16) gives

$$\mathbf{Q} = \mathbf{L}_{21} \mathbf{L}_{11}^{-1} \mathbf{J} \tag{1.19}$$

The tensor relating $\mathbf{Q}$ and $\mathbf{J}$ under these conditions is just the Peltier tensor Π. If we now apply the Onsager relation, Eq. (1.13), we immediately obtain the second Kelvin relation in its general form:

$$\Pi(\mathbf{H}) = T \check{\mathbf{S}}(-\mathbf{H}) \tag{1.20}$$

The derivation of the Kelvin relation illustrates only one of the many useful applications of the theory of irreversible thermodynamics. By a similar technique one can also derive relations among the galvano- and thermomagnetic coefficients [48C1, 52C1].

Using the four variables in Eqs. (1.15) and (1.16), there are a total of twelve relations of the type characterized by Eqs. (1.17) and (1.18). These are given in Table 1.3 [74T1]. The subscripts specify which quantity is maintained zero during the measurement. In this table the last column gives the value of the quantity when a relaxation time can be assumed, that is, when the scattering of the electrons is elastic and the Wiedemann–Franz law applies (see Chapter 2).

As Table 1.3 suggests, there are other thermoelectric parameters in addition to S, Π, and μ that could conceivably be of practical interest, although it is clear that no additional fundamental information can be gleaned from their measurement that could not be obtained from the more conventional ones. However, there may be important practical considerations, to which we address ourselves in Chapter 3, which could make one type of measurement more profitable. One in particular, namely, Π_e, is most attractive especially at low temperatures, and has in fact been measured in tungsten [74G1, 74G2] and silver [74T2]. The attractiveness of Π_e as contrasted with S is that the former should approach a constant value as T goes to zero, whereas S must vanish as T approaches zero. Moreover, there are practical considerations which allow for a more precise determination of Π_e as compared to S at low temperatures.

In concluding this introductory chapter, we direct the reader to two recent sources on thermopower, namely, the book by Barnard [72B2] and the review article by Huebener [72H1].

2 | Survey of the Theory of Electronic Conduction in Metals

In Chapter 1 we made no mention of the detailed mechanisms that lead to electronic conduction and associated thermoelectric effects. The general phenomenological equations (1.15) and (1.16) are equally valid for metals, semiconductors, or an ionized gas, and Onsager's relations are correct for any of these. This book is concerned, however, only with metallic conductors, and our interest in thermoelectric phenomena is based on the belief that they can shed light on fundamental features of electronic energy levels and on the interaction of conduction electrons with their environment. To see this more clearly, we must now review briefly the elements of transport theory as it applies to these materials.

2.1 Electrons in Metals

A metal, even an ideal pure single crystal, is an extremely complicated many-body system. The valence electrons, which in the metallic solid constitute the electrons in the conduction band(s), interact with each other through Coulomb and exchange forces and interact also with the ions of the crystal lattice. The latter, in turn, interact with each other through forces whose strength and spatial dependence are strongly influenced by the core as well as by the conduction electrons.

A very much simplified model of the perfect metallic crystal is therefore needed. Such a model is a regular array of positive ions embedded in a "gas" of electrons whose average charge density is such as to maintain macroscopic electrical neutrality. In this model we view the electron gas as an assembly of noninteracting charged particles of spin one-half, obeying Fermi–Dirac statistics. As regards the lattice of positive ions, the dominant features here are the spatial periodicity and point symmetry of the lattice (cubic, hexagonal, etc.). The ions, however, are not stationary but oscillate about their equilibrium positions; in the harmonic approximation the convenient description of lattice vibrations is in terms of *phonons*, that is, normal modes that correspond to plane waves, each characterized by wave vector $\mathbf{q}$, polarization index j, angular frequency ω, and energy $\hbar\omega$.

The conduction electrons feel the presence of the ions through their mutual Coulomb interaction. This interaction may be separated into two terms: one time-dependent, and the other time-independent. The time-dependent part of the interaction arises from the oscillatory motion mentioned above and leads to scattering of electrons due to absorption or emission of phonons. The time-independent, spatially periodic part is of fundamental importance since it gives rise to the well-known *Brillouin zone* structure and the appearance of electronic energy bands in solids [71K1].

In the following pages we shall first review the relevant results for a free electron gas. Next, we shall briefly indicate the modifications that arise as a result of the periodic lattice potential. Then, we shall review the derivation of the transport coefficients from the solution of the *Boltzmann equation* and, thereafter, turn our attention to the primary topic of interest— thermoelectricity.

2.1a *Free Electron Gas*

The energy levels of a gas of noninteracting free electrons, confined to a volume V, are given by

$$\varepsilon(\mathbf{k}) = \hbar^2 \mathbf{k}^2 / 2m \tag{2.1}$$

where $\mathbf{k}$ is the wave vector of the plane wave eigenfunctions

$$\psi_{\mathbf{k}}(\mathbf{r}) = \frac{1}{\sqrt{V}} e^{i\mathbf{k}\cdot\mathbf{r}} \tag{2.2}$$

and is related to the electron's momentum $\mathbf{p}$ by

$$\mathbf{p} = \hbar\mathbf{k} \qquad (2.3)$$

The density of states of this electron gas, defined as the number of energy levels per unit energy range per unit volume, is

$$N(\varepsilon) = \frac{(2m)^{3/2}}{2\pi^2\hbar^3}\varepsilon^{1/2} \qquad (2.4)$$

where we have taken the twofold spin degeneracy into account.

At equilibrium the distribution of electrons among the available energy levels is given by the *Fermi–Dirac distribution*

$$f_0(\varepsilon) = [1 + e^{(\varepsilon-\eta)/kT}]^{-1} \qquad (2.5)$$

The Fermi energy η is obtained from the expression for the electron density n_0 as follows. Taking as our zero of energy that of the lowest energy level, corresponding to the state $\mathbf{k} = 0$, we have

$$n_0 = \int_0^\infty N(\varepsilon)f_0(\varepsilon)\,d\varepsilon = \frac{(2mkT)^{3/2}}{2\pi^2\hbar^3}F_{1/2}(w) \qquad (2.6)$$

where $w = \eta/kT$, and $F_\nu(w)$ is the *Fermi–Dirac function* of order ν:

$$F_\nu(w) = \int_0^\infty \frac{x^\nu}{1+e^{x-w}}\,dx \qquad (2.7)$$

Electron densities in metals are generally so large that, at least below their melting points, $\eta \gg kT$ and $F_{1/2}(w)$ may be closely approximated by the first few terms of a series expansion. In this *degenerate limit* the integral

$$I = \int_0^\infty \frac{dg}{d\varepsilon}f_0(\varepsilon)\,d\varepsilon = g(\eta) - g(0) + 2\sum_{n=1}^\infty C_{2n}(kT)^{2n}\left(\frac{d^{2n}g}{d\varepsilon^{2n}}\right)_{\varepsilon=\eta} \qquad (2.8)$$

where g is an arbitrary, well-behaved (i.e., differentiable) function of ε, $C_{2n} = (1 - 2^{1-2n})\zeta(2n)$, and $\zeta(x)$ is the Riemann zeta function. The first two coefficients of the series are $C_2 = \pi^2/12$ and $C_4 = 7\pi^4/720$. Thus,

$$n_0 \simeq \frac{(2m)^{3/2}}{3\pi^2\hbar^3}\eta^{3/2}\left[1 + \frac{\pi^2}{8}\left(\frac{kT}{\eta}\right)^2 + \cdots\right] \qquad (2.9)$$

and

$$\eta \simeq \eta_0\left[1 - \frac{\pi^2}{12}\left(\frac{kT}{\eta_0}\right)^2 + \cdots\right] \qquad (2.10)$$

where η_0, the Fermi energy at $T = 0$ K, is given by

$$\eta_0 = \frac{\pi^2 \hbar^2}{2m} \left(\frac{3n_0}{\pi} \right)^{2/3} \tag{2.11}$$

2.1b   Energy Bands

The surface of constant energy in **k**-space for which $\varepsilon(\mathbf{k}) = \eta$ is known as the *Fermi surface*. Whereas for a free electron gas the Fermi surface is a sphere, that of a real metal is generally far more complicated, often displaying intricate topological features. A first-order, rough approximation to the real Fermi surface for simple polyvalent metals can be obtained by means of the *Harrison construction*, which is described in all texts on solid state physics [71K1].

As suggested above, the periodic potential of the lattice modifies the energy-level structure as well as the eigenfunctions. This potential gives rise to the well-known Brillouin zone structure, these zones in **k**-space being defined by the polyhedra formed from the planes which bisect the vectors $\mathbf{K}_n$ of the *reciprocal lattice* [71K1]. A general reciprocal lattice vector is $\mathbf{K}_n = n_1 \mathbf{K}_1 + n_2 \mathbf{K}_2 + n_3 \mathbf{K}_3$, where n_1, n_2, n_3 are integers, and the primitive vectors of the reciprocal lattice are determined from those of the real lattice $\mathbf{a}_1, \mathbf{a}_2, \mathbf{a}_3$, by the relation

$$\mathbf{K}_1 = 2\pi \frac{\mathbf{a}_2 \times \mathbf{a}_3}{\mathbf{a}_1 \cdot (\mathbf{a}_2 \times \mathbf{a}_3)} \tag{2.12}$$

One can easily show that each Brillouin zone can, in a monatomic Bravais lattice, accommodate an electron density corresponding to two electrons per atom.

At **k** vectors which terminate on a Brillouin zone boundary, the energy eigenvalues exhibit a discontinuity, and, along a given direction in **k**-space, the energy spectrum of electrons in solids displays the well-known pattern of energy bands separated by forbidden energy gaps. Since these forbidden gaps occur at different energies for different directions in **k**-space, overlapping energy bands are frequently encountered. The formal description of the energy bands is usually achieved in the framework of the *reduced zone scheme* or of the *periodically extended zone scheme*. Which of these is used is largely a matter of convenience [60Z1].

The eigenfunctions of electrons in a crystal are no longer plane waves but are the *Bloch waves*

$$\psi_{\mathbf{k}}(\mathbf{r}) = u_{\mathbf{k}}(\mathbf{r}) e^{i\mathbf{k} \cdot \mathbf{r}} \tag{2.13}$$

Here $u_k(\mathbf{r})$ is a function which has the periodicity of the lattice. Since these functions are no longer eigenfunctions of the momentum operator, the quantity $\hbar k$ is referred to as the *crystal momentum* of the electron, partly because under the influence of an external driving force $\mathbf{F}$ the wave vector $\mathbf{k}$ changes with time according to

$$\hbar \dot{\mathbf{k}} = \mathbf{F} \tag{2.14}$$

The velocity of an electron in the state $\mathbf{k}$ is not $\hbar k/m$ but is given by the expression

$$\mathbf{v}(\mathbf{k}) = \frac{1}{\hbar} \nabla_k \varepsilon(\mathbf{k}) \tag{2.15}$$

Finally, the general expression for the density of states is

$$N(\varepsilon) = \frac{1}{4\pi^3} \int_\varepsilon \frac{d\mathscr{S}}{|\nabla_k \varepsilon|} = \frac{1}{4\pi^3} \int_\varepsilon \frac{d\mathscr{S}}{\hbar |\mathbf{v}(\mathbf{k})|} \tag{2.16}$$

where $d\mathscr{S}$ is the surface element in $\mathbf{k}$-space, and the integral is over the surface of constant energy ε.

2.2 Transport Properties

Expressions for the transport coefficients are derived by solving the Boltzmann transport equation subject to appropriate boundary conditions. Under steady-state conditions in a bulk sample the distribution function must satisfy the equation

$$\dot{\mathbf{k}} \cdot \nabla_k f + \mathbf{v}_k \cdot \nabla_r f = \left(\frac{\partial f}{\partial t} \right)_c \tag{2.17}$$

where f is a function of $\mathbf{k}$ and $\mathbf{r}$. The terms on the left-hand side of Eq. (2.17) are the *drift* terms, which describe the change in the local distribution due to a time rate of change of $\mathbf{k}$ and $\mathbf{r}$ (acceleration and velocity). The term on the right-hand side is the *collision* term, which gives the number of electrons scattered into the six-dimensional volume element $d\mathbf{k}\,d\mathbf{r}$ per unit time. The calculation of the collision term often presents some of the more difficult problems, and the result depends critically on the type of interactions which give rise to scattering, e.g., interaction with stationary defects, lattice vibrations, mutual exchange interaction, etc. In many situations, however, the collision term can be approximated by means of a *relaxation time* τ [68B1], i.e.,

$$\left(\frac{\partial f}{\partial t}\right)_c = -\frac{f-f_0}{\tau} = -\frac{f_1}{\tau} \tag{2.18}$$

In the simplest cases τ is assumed independent of the direction of $\mathbf{k}_i$, where $\mathbf{k}_i$ is the initial wave vector, that is, the wave vector of the electron before it is scattered, although τ will generally be a function of electron energy. In the real metal, however, the relaxation time will always be anisotropic over a constant energy surface, and calculations based on the assumption of isotropy must, therefore, be viewed with some skepticism.

With the use of the linearization approximation $\nabla f \simeq \nabla f_0$, the drift term can be rewritten

$$[e\mathbf{E} - (\varepsilon - \eta)\nabla \ln T] \cdot \mathbf{v}_k \frac{\partial f_0}{\partial \varepsilon} - \frac{e}{\hbar^2 c}\left[\mathbf{H} \cdot \mathbf{\Omega}(f_1)\right] \tag{2.19}$$

where $\mathbf{\Omega}$ is the operator

$$\mathbf{\Omega} = \nabla_k \varepsilon \times \nabla_k \tag{2.20}$$

Most of the material in this book will be concerned with thermoelectric effects in the absence of a magnetic field, and we shall therefore neglect the last term in (2.19) hereafter. In any event, in one of the more interesting experimental situations in which the magnetic field is sufficiently strong to permit observation of orbital quantization of the electronic energy levels, leading to de Haas–van Alphen type oscillations of S and Π, the above formulation is quite inappropriate.

The electric and heat current densities are given by

$$\mathbf{J} = e \int \mathbf{v}_k f_1(\mathbf{k}) \, d\mathbf{k};$$

$$\mathbf{Q} = \int \mathbf{v}_k (\varepsilon_k - \eta) f_1(\mathbf{k}) \, d\mathbf{k} \tag{2.21}$$

where we have replaced f by f_1, since under equilibrium conditions $\mathbf{J}$ and $\mathbf{Q}$ must vanish. Equation (2.21) can now be rewritten in the form of (1.10), where

$$\mathscr{L}_{11} = e^2 \mathbf{K}_0$$

$$\mathscr{L}_{12} = -\frac{1}{T}\mathbf{L}_{21} = -\frac{e}{T}\mathbf{K}_1 \tag{2.22}$$

$$\mathscr{L}_{22} = -\frac{1}{T}\mathbf{K}_2$$

and in the relaxation time approximation

$$\mathbf{K}_n = -\int \mathbf{v}_k \mathbf{v}_k \tau(\mathbf{k})(\varepsilon - \eta)^n \frac{\partial f_0}{\partial \varepsilon} \, d\mathbf{k}$$

$$= -\frac{1}{4\pi^3} \int\!\!\int \mathbf{v}_k \mathbf{v}_k \tau(\mathbf{k})(\varepsilon - \eta)^n \frac{1}{|\nabla_k \varepsilon|} \frac{\partial f_0}{\partial \varepsilon} \, d\mathcal{S} \, d\varepsilon \qquad (2.23)$$

Accordingly, the transport parameters, such as the isothermal electrical conductivity σ_i, the electronic thermal conductivity κ_i, and thermoelectric power $\mathbf{S}$ (see Table 1.3), can be expressed in terms of the above integrals as follows:

$$\sigma_i = e^2 \mathbf{K}_0 \qquad (2.24a)$$

$$\kappa_s = \frac{1}{T}(\mathbf{K}_2 - \mathbf{K}_1 \mathbf{K}_1 \mathbf{K}_0^{-1}) \qquad (2.24b)$$

$$\mathbf{S} = \frac{1}{eT} \mathbf{K}_1 \mathbf{K}_0^{-1} \qquad (2.24c)$$

In the degenerate limit, expansion of the the integrals (2.23) using (2.8) leads to the Mott expression

$$S_d = \frac{\pi^2}{3} \frac{k^2 T}{e} \left(\frac{\partial \ln \sigma}{\partial \varepsilon}\right)_\eta \qquad (2.25)$$

Two points need to be stressed here. First, we have added a subscript d to indicate that we are concerned here only with what is now generally called the *diffusion thermoelectric power*. This is the contribution to the TEP associated with a system of electrons that interact with a random distribution of scattering centers *that are assumed to exist in thermal equilibrium at the local temperature T.* As we shall see presently, this assumption is a very poor approximation in the real situation, and an additional contribution to the TEP appears, the *phonon-drag thermopower*, when the assumption of local thermal equilibrium is lifted.

Second, although Eq. (2.25) has been derived here with the aid of the relaxation time approximation, the result is generally valid, provided the conduction electrons constitute a degenerate Fermi gas, so that higher-order terms in the expansion (2.8) are negligible.

2.3 Relaxation Time Anisotropy for Spherical Fermi Surfaces

Having previously commented on the matter of relaxation time anisotropy, we digress here to show how anisotropy may, and generally does, arise even when the Fermi surface is spherical. The quantity $\sigma(\varepsilon)$ appearing in Eq. (2.25) is, of course, an average over a surface of constant energy. It is quite possible that different portions of that surface may contribute unequally to $\sigma(\varepsilon)$, and that the manner in which this anisotropy arises may depend on the source of the scattering interaction, e.g., electron–phonon or electron–impurity scattering. Such differences in anisotropy can then have a profound effect on the changes in the transport properties that appear when impurities are added to the nominally pure material.

In an electron–phonon scattering event, the wave vector of the electron changes from $\mathbf{k}_i$ to $\mathbf{k}_f$ through the absorption or emission of a phonon of wave vector $\mathbf{q}$. In such a process crystal momentum is conserved, that is,

$$\mathbf{k}_f = \mathbf{k}_i \pm \mathbf{q} + \mathbf{K} \qquad (2.26)$$

where the plus sign corresponds to absorption and the minus sign to emission of a phonon of wave vector $\mathbf{q}$, and $\mathbf{K}$ is a reciprocal lattice vector (which may be zero). If $\mathbf{K} = 0$, the process is called a *Normal* (N) event; if $\mathbf{K} \neq 0$, it is called an *Umklapp* (U) event. Coupling parameters for N and U processes are generally different, and different branches of the vibrational spectrum contribute differently to N and U events.

In addition to crystal momentum conservation, energy must also be conserved in the scattering process, that is,

$$\varepsilon(\mathbf{k}_f) = \varepsilon(\mathbf{k}_i) \pm \hbar\omega_q \qquad (2.27)$$

where $\hbar\omega_q$ is the energy of the phonon that is either absorbed ($+$) or emitted ($-$). In many cases it is a fairly good approximation to neglect the small energy change of the electron and to treat the scattering process as quasi-elastic, and in the following we shall make this simplifying approximation.

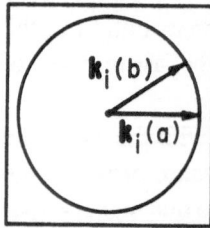

Fig. 2.1. A square Brillouin zone with a circular Fermi "surface." Two initial electron states $k_i(a)$ and $k_i(b)$ are shown.

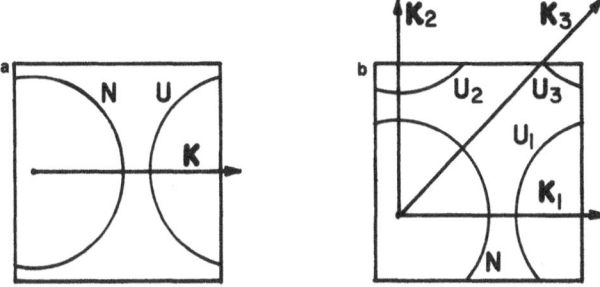

Fig. 2.2. Possible N and U scattering loci on the Fermi surface of
Fig. 2.1 for initial electron state $k_i(a)$, (a), and $k_i(b)$, (b).

Thus, an electron–phonon scattering event takes an electron from a given point on the Fermi surface to another point, again on the Fermi surface. Let us consider then a two-dimensional square lattice, with a square Brillouin zone and a circular Fermi "surface" as shown in Fig. 2.1. All of the allowed phonon wave vectors are, of course, also contained within the same square Brillouin zone. We now ask the following question: Given an initial electron state $\mathbf{k}_i$, what is the locus of phonons that can scatter this electron to another point on the Fermi surface? The answer to that question is obtained by a simple geometrical construction in which the Fermi surface is displaced by the distance $-\mathbf{k}_i$ in the Brillouin zone [74G3]. Figure 2.2 illustrates two cases: when $\mathbf{k}_i$ is in the $(1, 0)$ direction (a), and when $\mathbf{k}_i$ is near the $(1, 1)$ direction (b). Phonons on the portions of the circle marked N and U participate in N and U processes, respectively. Evidently, a completely different group of phonons induce transitions of an electron with initial wave vector near $(1, 0)$ and near $(1, 1)$, and it therefore follows that even though the Fermi surface is isotropic, the relaxation time $\tau(\mathbf{k})$ due to electron–phonon scattering will show marked anisotropy.

2.4 Thermopower: Isotropic Relaxation Time Approximation

Returning now to the discussion of S_d, we see that according to Eq. (2.25) the sign of S_d will be determined by that of the charge carriers and the sign of the logarithmic derivative of $\sigma(\varepsilon)$. From (2.23) and (2.24) we can

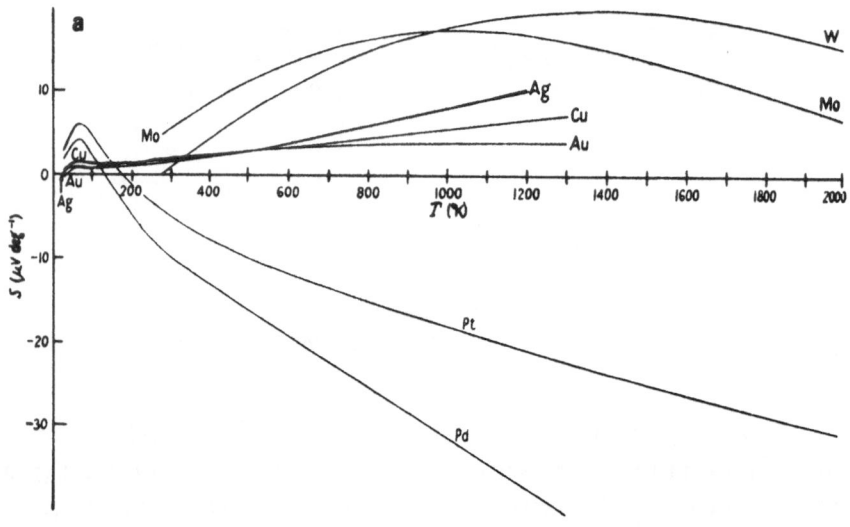

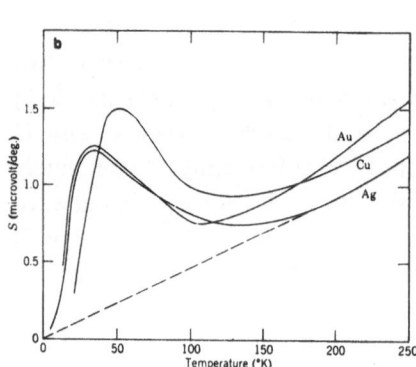

Fig. 2.3. The thermoelectric power of metals.
(a) Refractory and noble metals over a wide
temperature range (from [58C2]); (b) copper,
silver, and gold between 0 and 250°K (from
[62M1]); (c) alkali metals at low temperatures
(from [62M1]).

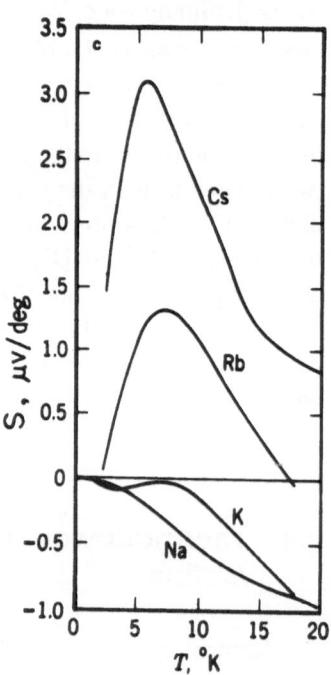

write

$$\sigma(\varepsilon) = \frac{e^2 l(\varepsilon) \mathcal{S}(\varepsilon)}{12\pi^3 \hbar} \tag{2.28}$$

where

$$l(\varepsilon) = \tau(\varepsilon)v(\varepsilon) \tag{2.29}$$

is the mean free path of electrons of energy ε, averaged over the surface of constant energy ε, and $\mathcal{S}(\varepsilon)$ is the area of this surface in k-space.

Provided the energy dependence of $\tau(\varepsilon)$ does not vary with temperature (though, of course, its magnitude will), S_d should be a linear function of temperature. This behavior appears to be obeyed, at least approximately, by the measured thermopowers of nontransition metals at temperatures in excess of θ_D, where θ_D is the *Debye temperature* and also at very low temperatures. However, in the region between about $\theta_D/100$ and θ_D marked departures from the linear dependence are generally observed (see Fig. 2.3). In most instances, these departures are attributable to phonon drag. We shall return to this matter presently, but first we shall consider the diffusion term further.

In pure metals at temperatures above a few degrees Kelvin, the relaxation time is limited primarily by electron–phonon scattering. We now know that this will lead to an anisotropic $\tau(k)$, and that, consequently, different portions of the Fermi surface may have relaxation times of different magnitudes and energy variation. Still, elementary calculations [53W1, 60Z1] predict that $\tau(\varepsilon) \propto \varepsilon^{3/2}$, and as a first approximation one may ascribe this energy dependence to the average relaxation time. Since for a spherical surface $v(\varepsilon) \propto \varepsilon^{1/2}$ and $\mathcal{S}(\varepsilon) \propto \varepsilon$, the logarithmic derivative of (2.25) takes the value $3/\eta$. On the other hand, at very low temperatures, and in dilute alloys also at somewhat higher temperatures, $\tau(\varepsilon)$ is limited by impurity scattering. The energy dependence of τ in this *residual resistance* region is generally $\varepsilon^{-1/2}$, and, consequently, the logarithmic derivative is $1/\eta$. Hence, the diffusion thermopower in the free-electron approximation is

$$S_d = \frac{\pi^2 k^2 T}{3e\eta} \qquad \text{very low temperature (residual resistance region)}$$

$$S_d = \frac{\pi^2 k^2 T}{e\eta} \qquad \text{high temperature (phonon scattering region)}$$

Detailed calculations, based on the variational formulation of transport, show that the two linear dependences join smoothly in the intermediate region [53W1].

2.5 Thermopower: Real Metals

2.5a *Alkali and Noble Metals*

Since the electronic charge e is negative, the sign of S_d for a free-electron gas under these scattering conditions must also be negative. The positive thermoelectric power of lithium and of the noble metals copper, silver, and gold at temperatures at which phonon drag is unquestionably very small has been a perplexing problem for some years.

The Fermi surface of real metals is, of course, generally far from spherical. In the monovalent noble metals the Fermi surface is known to contact the Brillouin zone boundaries as shown in Fig. 2.4. In that case $\partial \mathscr{S}/\partial \varepsilon$ may well be negative. Nevertheless, it is now generally believed that Fermi surface anisotropy by itself is probably not the principal reason for the positive diffusion thermopowers of these monovalent metals. First, a Fermi surface which is of such predominantly holelike character as to yield a positive S_d would also generally result in a positive Hall coefficient; yet, all of the monovalent metals have negative Hall coefficients of roughly the correct magnitude appropriate to a nearly free electron gas. Second, the thermopowers of these metals remain positive in the liquid state where, presum-

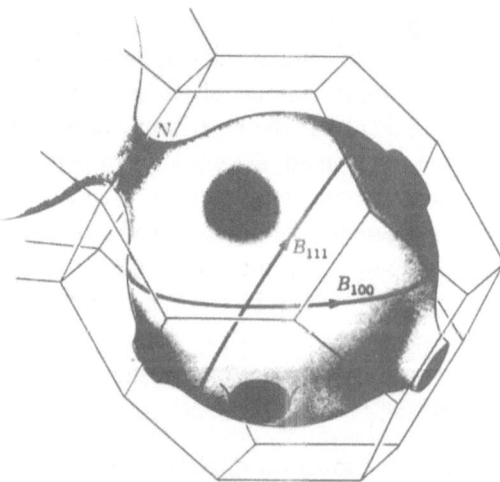

Fig. 2.4. The Fermi surface of copper. N, B_{111}, and B_{100} indicate three extremal electron orbits observed in de Haas–van Alphen oscillations (from [71K1]).

ably, Fermi surface effects should be minimal. Hence, the cause for the positive S_d is thought to arise from a strongly negative value of $[\partial \ln l(\varepsilon)/\partial \varepsilon]_\eta$, which could be obtained even under the assumption of a spherical Fermi surface.

Before we outline the appropriate calculation [67R1, 68R1], it is worthwhile to gain some physical insight into the mechanism by which, even for free electrons, S_d could become positive. In essence the conceptual difficulty relates to the fact that a positive S_d for negatively charged electrons implies that under the influence of a temperature gradient the diffusive motion of these particles is *up* the temperature gradient rather than down.

From Eqs. (2.18), (2.19), and (2.29) we can write for the solution for $f_1(\mathbf{k})$

$$f_1(\mathbf{k}) = -\frac{\partial f_0}{\partial \varepsilon}\left[e\mathbf{E}\cdot\mathbf{l}(\mathbf{k}) - \frac{\varepsilon_k - \eta}{T}\nabla T \cdot \mathbf{l}(\mathbf{k})\right] \tag{2.30}$$

In (2.30) the quantity within the brackets is just the work done on a carrier of wave vector $\mathbf{k}$ during one mean free path $\mathbf{l}(\mathbf{k})$. Of crucial importance in understanding diffusion thermopower is the fact that the entropy change $[\varepsilon(\mathbf{k}) - \eta]/T$ reverses sign as $\varepsilon(\mathbf{k})$ passes through the Fermi energy η. Thus, the effective "thermal" or "statistical" force also changes sign at the Fermi energy. The direction of the net diffusive motion for the entire distribution of electrons is, therefore, the result of a competitive balance between carriers above and below the Fermi energy, in which those carriers with the longer mean free path dominate. There is nothing in the solution of the transport equation that requires this diffusion to be either parallel or antiparallel to ∇T. The energy dependence of $\mathscr{S}(\varepsilon)$ now appears principally as a statistical factor which determines the phase space available for excitation of the free-electron gas from its equilibrium configuration in $\mathbf{k}$-space. Thus, the critical parameter appears to be the energy-dependent mean free path $\mathbf{l}(\mathbf{k})$.

The electronic mean free path for lattice scattering is given by

$$\frac{1}{\mathbf{l}(\mathbf{k})} = \frac{m^2 \Omega_0}{4\pi\hbar^4 k^4}\int_0^{2k} q^3 |V(\mathbf{q})|^2 S(\mathbf{k}, \mathbf{q})\, dq \tag{2.31}$$

where

$$S(\mathbf{k}, \mathbf{q}) \equiv \frac{1}{2\pi}\int S'(\mathbf{q})\, \delta[\mathbf{q}\cdot\mathbf{k} + (q/2k)]\, d\Omega(\mathbf{q}) \tag{2.32}$$

Here $S'(\mathbf{q})$ is the static structure factor of the phonons, Ω_0 is the volume of the unit cell, $d\Omega(\mathbf{q})$ is the element of solid angle about $\mathbf{q}$, and $V(\mathbf{q})$ is the qth Fourier component of the spherically symmetric local pseudopotential.

For the noble metals and also for lithium, $V(\mathbf{q})$ is negative for small q, and crosses the axis and becomes positive for q less than $2k_F$, where k_F is the Fermi wave vector. Thus, for large values of q, $\partial|V(q)|^2/\partial q$ is positive. As Robinson has shown, $S(\mathbf{k}, \mathbf{q})$ is a monotonically increasing function of $q/2k_F$, which, along certain crystallographic directions, increases quite sharply as $q/2k_F$ approaches unity. Hence, the "anomalous" dependence of $|V(q)|^2$ on q for large momentum transfer is enhanced by the large structure factor in this region. Consequently, $l(k)$ decreases quite drastically with increasing k, i.e., with increasing electron energy, and this dependence gives rise to the large negative value of $[\partial \ln l(\varepsilon)/\partial \varepsilon]_\eta$. In Robinson's calculations a free-electron gas is assumed, and all of the mean free path anisotropy, which can be quite substantial, arises from $S(\mathbf{k}, \mathbf{q})$. Even a very rough averaging procedure yields values of $[\partial \ln l(\varepsilon)/\partial \varepsilon]_\eta$ which agree reasonably well for the alkali metals. In the case of the noble metals, Robinson has presented plausibility arguments to show that for these metals also the positive diffusion thermopower can be reconciled with a free-electron model.

Hasegawa and Kasuya [68H1], on the other hand, have examined the diffusion thermopower of copper using a very different approach. They rely not on the pseudopotential but on the Bardeen formula for the electron–phonon matrix element, and, in contrast to Robinson, calculate $v(\varepsilon)$ from detailed band structure calculations by Burdick [63B1]. They find that the term $(\partial \ln v/\partial \varepsilon)_\eta$ is substantially greater than in the free-electron approximation, and that this is largely responsible for the positive value of S_d of copper. Although they do obtain the correct sign, the calculated magnitude of S_d is too small by a factor of four. It is possible that a calculation which combines the two approaches might well give a satisfactory account of the diffusion thermopower of the noble metals.

2.5b Polyvalent Metals

In the case of polyvalent metals the situation is so complicated that calculations along the above lines have not been attempted. For these metals the Fermi surface consists of several sheets, some simply and some multiply connected, with both electron and hole character. Although the Fermi surface configuration for most metals is now well known from de Haas–van

Alphen and similar experiments, the dependence of $v(k)$ on k (both magnitude and direction) is not so well established. Moreover, interband or intersheet scattering is another feature that further compounds the complications in performing detailed calculations. Generally, for these metals theory has progressed little beyond the old multiband model in which it is assumed that the very complex Fermi surface may be approximated by two or more spherical surfaces with electron and/or hole character. Under these assumptions the total diffusion thermopower arising from the simultaneous action of these "bands" is given by

$$S_d = \sum_j \sigma_j S_d^j \Big/ \sum_j \sigma_j \qquad (2.33)$$

where the j subscript labels the band, and S_d^j is the diffusion thermopower that one would calculate if only the jth band contributed.

In all cases, an important factor determining the sign and magnitude of S_d is the energy dependence of $\tau(\varepsilon)$. If, moreover, the functional form of $\tau(\varepsilon)$ depends on temperature, S_d will show temperature variations that deviate from the linear form predicted by Eq. (2.25). This situation can arise even in a "one-band" model whenever two or more relaxation mechanisms are of comparable importance but have different temperature dependences. Two cases will be discussed in later chapters. The first concerns the influence of electron–electron scattering on S_d in transition metals, and the second relates to the thermopower anomaly associated with the resistance minimum in some dilute alloys containing transition-metal solutes, i.e., the *Kondo effect.* Similarly, nonlinear behavior of S_d with temperature can arise in a multiband model if the relative conductivities associated with the several bands change with temperature.

Recent work by Nielsen and Taylor [68N1, 70N1, 70N2, 74N1] has shown that the diffusion thermopower may display a temperature dependence at variance with the preceding predictions even in a simple free-electron approximation. The theory considers the influence of intrinsic two-phonon processes on the electron relaxation time in a second-order treatment of electron–phonon scattering. It is found that, although the inclusion of two-phonon events does not affect the relaxation time significantly, it does have a profound influence on the energy dependence of $\tau(\varepsilon)$ just at the Fermi energy. Moreover, this functional dependence changes with temperature, resulting in a contribution to the TEP that varies as T^3 at low temperatures and as T^{-1} at higher temperatures, attaining a maximum in the vicinity of $0.4\theta_D$, where θ_D is the Debye temperature. As we shall see presently, this is just the temperature dependence that is expected from the phonon-drag contribution, and we shall, therefore, include a more detailed discussion of this "phony phonon drag" in Chapter 4. However, the origin of this additional term is strictly in S_d; indeed, in the theory of Nielsen and Taylor the phonon system is assumed to be in local thermal equilibrium. One of the perplexing problems facing workers in the field is that of distinguishing between true and phony phonon drag. At this time there appears to be no unequivocal method for assessing the relative importance of the two effects, which both lead to the same behavior for the measured thermopower.

2.6 Phonon Drag

The peak in the thermoelectric power of the noble metals near 60 K is the typical pattern found in monovalent metals. Polyvalent metals frequently exhibit a more complicated behavior, as shown in Fig. 2.5, and in noncubic metals the thermopower, especially at low temperatures, is generally a sensitive function of the orientation of ∇T relative to the crystallographic axes (Fig. 2.5).

Numerous experimental studies have suggested that these departures from the approximately linear dependence of S on T predicted by Eq. (2.25) are the result of phonon drag. A detailed discussion of phonon drag is deferred to Chapter 4. We shall present here only a brief qualitative account of this phenomenon.

In any crystalline solid a thermal gradient will induce heat transport via lattice vibrations. Hence, under these conditions not only the electron but also the phonon distribution will depart from equilibrium, and the temperature gradient will establish a net phonon current traveling from the hot to the cold end of the sample. The mean free path l_g of these phonons is limited by a variety of interactions, such as phonon–phonon, phonon–electron, and phonon–imperfection scattering. The average mean free path may be deduced from the lattice thermal conductivity κ_g through the kinetic theory relation

$$\kappa_g = \tfrac{1}{3}C_g u l_g \tag{2.34}$$

where C_g is the lattice specific heat, and u is the phonon velocity (approximately the velocity of sound).

To simplify matters and bring the role of phonon–electron interaction in thermoelectricity into focus, we assume for the moment that all other

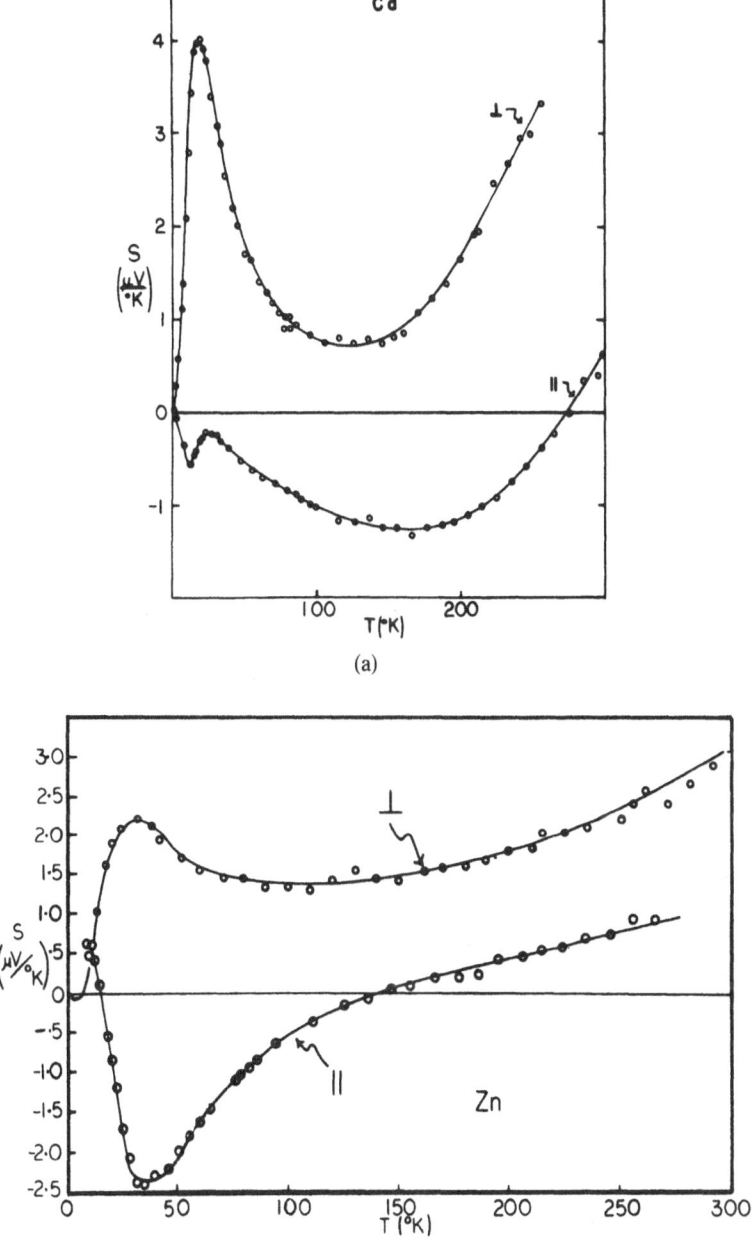

(a)

(b)

Fig. 2.5. The thermopower of Cd and Zn parallel and perpendicular to the hexagonal axis (from [70R1]).

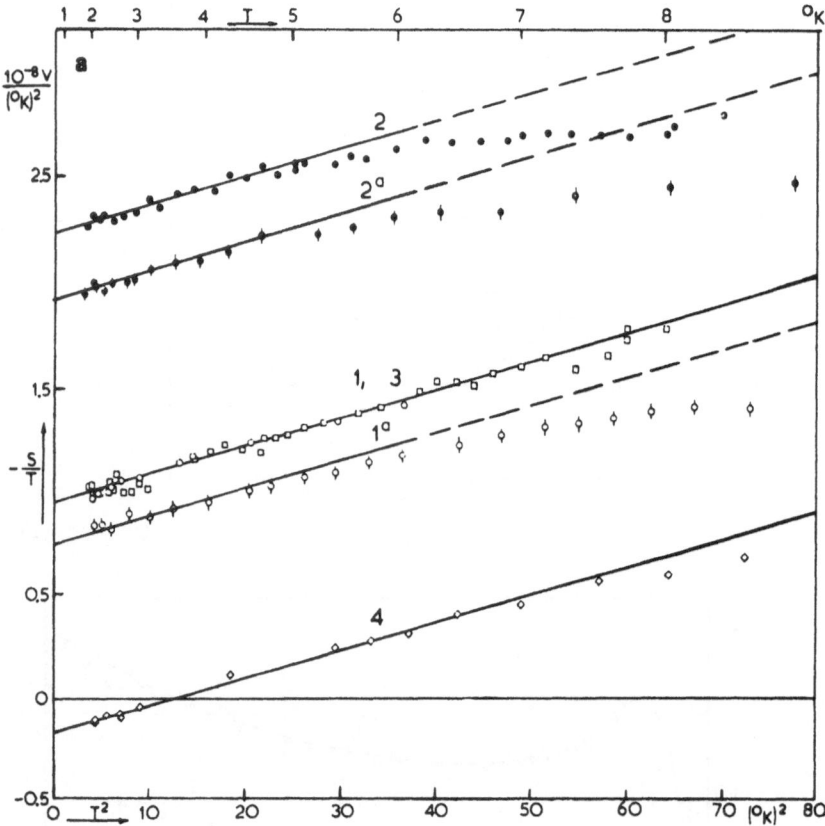

Fig. 2.6. The thermopower of metals at low temperature. Here S/T is plotted as a function of T^2. (a) Six aluminum samples: samples 1 and 2 are "pure" Al, annealed; 1a and 2a are Al, strained at 20°C; sample 3 is an alloy containing 1.2% Mg; sample 4 contains 0.89% Mg (from [60D1]). (b) Three tin–indium alloys; (2) 0.87% In, (3) 0.51% In, (7) 0.52% In; all samples are single crystals and (3) and (7) had different orientations (see Fig. 2.5) (from [64V1]). (c) Pure indium (from [74S1]).

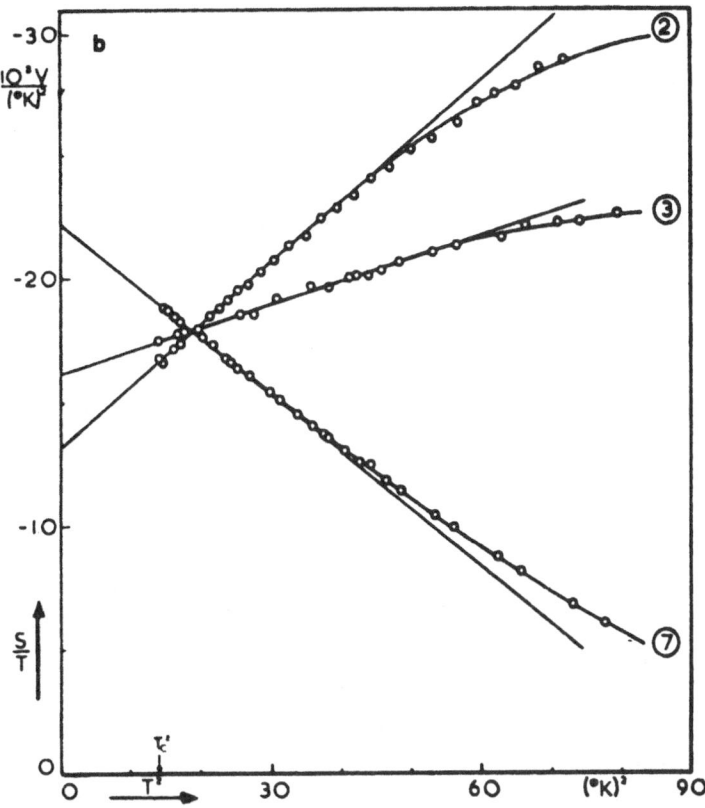

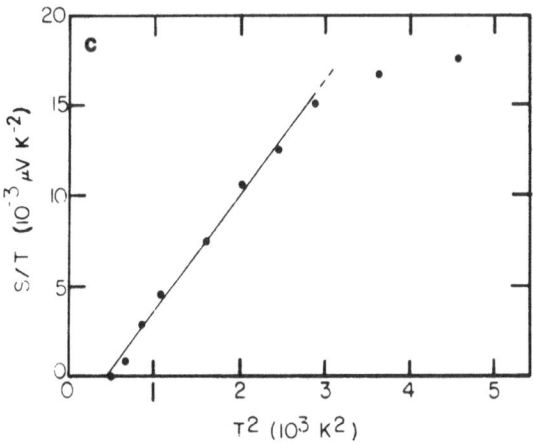

phonon scattering processes can be neglected. In each phonon–electron scattering event the crystal momentum and energy conservation rules, (2.26) and (2.27), are obeyed. As we shall see presently, N and U processes can lead to quite different contributions to phonon drag and have to be considered fairly carefully. However, for the moment we shall neglect U processes altogether.

As a result of the phonon current generated by the temperature gradient, an electron within the metal will be more likely to absorb a phonon travelling in the $-\nabla T$ direction than one traveling in the opposite direction. Therefore, the electron distribution will tend to absorb the phonon momentum along $-\nabla T$ and will be "dragged" along by the phonon current as in viscous flow. Consequently, electrons tend to pile up at the cold end of the sample. This charge imbalance generates an internal electric field which exerts a retarding force on the streaming electrons, and ultimately a steady state is attained in which the total electric current vanishes, as required by the experimental boundary conditions. Clearly, phonon drag constitutes a new contribution to the thermoelectric field, supplementing that derived previously under the assumption that the scattering centers, in this case phonons, are in thermal equilibrium at the local temperature T.

The magnitude and temperature dependence of this phonon-drag thermopower, denoted by the symbol S_g, may be estimated as follows: Consider an acoustic wave propagating along the x direction through an isotropic medium containing charged, sound-absorbing particles. The radiation pressure associated with the spatial decay of acoustic energy U is $P_x = -(dU/dx)$. In steady state this force per unit area must be compensated by the opposing force of electrostatic origin associated with the above-mentioned charge imbalance, namely, $F_x = -n_0 e E_x$. If we now replace U by the energy density of thermal phonons propagating along one of the Cartesian coordinate directions, we have $dU/dx = (C_g/3)(dT/dx)$. Hence, the electrostatic field required to maintain zero current flow is

$$E_x = (C_g/3n_0 e)(dT/dx)$$

and hence the corresponding thermoelectric power is just

$$S_g = C_g/3n_0 e \tag{2.35}$$

To compare the relative magnitudes of S_g and S_d, it is convenient to rewrite (2.25) as

$$S_d = \frac{2C_e}{3n_0 e}\left(\frac{\partial \ln \sigma(\varepsilon)}{\partial \ln \varepsilon}\right)_\eta \tag{2.25a}$$

where C_e is the electronic specific heat. Evidently, S_g/S_d should be roughly equal to C_g/C_e. If we recall that nearly all the heat capacity of a metal at ordinary temperatures is due to excitation of phonons, we expect S_g to be much larger than S_d except at very low temperatures ($T \lesssim 2\text{–}5$ K). Moreover, we can also anticipate the low-temperature behavior of the total thermopower. As we raise the temperature from near 0 K, we expect that

$$S = S_d + S_g = a_e C_e + a_g C_g = AT + BT^3 \qquad T \ll \theta_D \qquad (2.36)$$

At high temperatures C_g should attain its classical value $3R$, and so we expect that S might be almost temperature independent since $C_e \ll C_g$.

Experimental results have qualitatively confirmed the low-temperature behavior given by Eq. (2.36), under conditions where this expression can be expected to hold, although generally $|B_{exp}| \ll |B_{calc}|$, and S_g may be of either sign (Fig. 2.6). At temperatures $T \gtrsim \theta_D/20$, however, the observed pattern departs significantly from that predicted by (2.36). Instead of approaching a constant value, S_g attains a maximum, usually at a temperature near $\theta_D/5$ (see Fig. 2.3). The reason for these departures is to be found in the phonon relaxation processes that we have previously ignored.

Although in a pure metal and at sufficiently low temperatures phonon–electron scattering is indeed the dominant phonon relaxation mechanism, at higher temperatures, as the vibrational amplitudes increase, phonon–phonon scattering due to the anharmonic terms in the lattice potential becomes increasingly important. The magnitude of S_g depends not merely on the deviation of the phonon spectrum from equilibrium, but also on the relative importance of phonon–electron scattering, compared to the sum total of all phonon relaxation processes, in restoring an equilibrium phonon distribution. Thus, in this approximation the phonon-drag thermopower is given by

$$S_g = \frac{C_g}{3n_0e}\left(\frac{\tau_p'}{\tau_p' + \tau_{pe}}\right) = S_g^0(T)\alpha(T) \qquad (2.37)$$

where τ_{pe} is the phonon relaxation time due to phonon–electron scattering, and τ_p' is the relaxation time due to all other phonon scattering processes.

We can now understand the general features of the high-temperature behavior of S_g. For $T \gtrsim \theta_D$, C_g is a constant and τ_{pe} is also independent of temperature. Since the dominant scattering mechanism, anharmonic phonon–phonon scattering, gives a relaxation time that is inversely proportional to temperature [58K1], we conclude that S_g should also be inversely proportional to T in this temperature region.

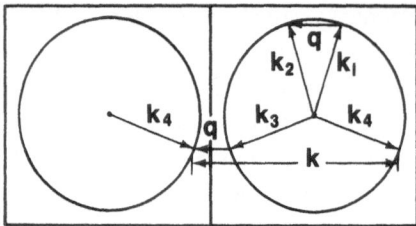

Fig. 2.7. N and U processes in the reduced and extended zone scheme. The transition $\mathbf{k}_1$ to $\mathbf{k}_2$ by absorption of the phonon $\mathbf{q}$ is an N process; note that $\Delta\mathbf{k} = \mathbf{k}_2 - \mathbf{k}_1$ is parallel to $\mathbf{q}$. The transition $\mathbf{k}_3$ to $\mathbf{k}_4$ by absorption of the phonon $\mathbf{q}$ is a U process; note that $\Delta\mathbf{k} = \mathbf{k}_4 - \mathbf{k}_3$ is antiparallel to $\mathbf{q}$.

The thermopower of most simple metals follows the behavior predicted by Eqs. (2.25) and (2.37) at least qualitatively. However, in all cases there are departures from these predictions that can be traced to the failure of this very elementary theory to take proper account of the true phonon spectrum, the electronic band structure, and details of the phonon–electron scattering process. A correct calculation of phonon drag must proceed along the following lines.

The contribution δS_g to S_g from a surface element $\delta\mathscr{S}$ of the Fermi surface due to the interaction of these electrons with phonons with polarization index j in the interval $d\mathbf{q}$ is calculated, using the appropriate electron–phonon coupling parameter. The evaluation of the total phonon–drag contribution then requires integration over the true phonon spectrum and over the Fermi surface. The formalism for performing such calculations has been developed by Bailyn [67B1, 60B1, 58B1], but rather than describe this in detail here, we shall consider in a qualitative manner how departures from the "ideal" in either the electronic or vibrational spectrum may influence S_g.

To illustrate the important difference between N and U processes, consider the situation illustrated in Fig. 2.7, which represents an ideal, i.e., spherical, Fermi surface. The transition from $\mathbf{k}_1$ to $\mathbf{k}_2$ through the absorption of a phonon of wave vector $\mathbf{q}$ changes the electron velocity by $\Delta\mathbf{v} = \mathbf{v}_2 - \mathbf{v}_1 = (\hbar/m)(\mathbf{k}_2 - \mathbf{k}_1) = \hbar\mathbf{q}/m$. Hence, it appears that through N scattering a net flow of phonons along the $-x$ direction causes a drift of electrons in this same direction, resulting in a negative thermopower if we recall the usual sign convention and the negative electronic charge.

The same phonon can also induce a U process, between electron states $\mathbf{k}_3$ and $\mathbf{k}_4$, involving the reciprocal lattice vector $\mathbf{K}$. In this case $\Delta\mathbf{v} = \mathbf{v}_4 - \mathbf{v}_3 = (\hbar/m)(\mathbf{k}_4 - \mathbf{k}_3) \neq \hbar\mathbf{q}/m$. In fact, the change in electron velocity is antiparallel to $\mathbf{q}$, and the net result of this U process is, therefore, a positive contribution to the phonon-drag thermopower. Although we have taken two very special cases, it is generally true that, when averaged over the Fermi surface and over the full phonon spectrum, the effect of N processes is a negative phonon-drag thermopower, whereas U processes contribute a positive phonon-drag thermopower to an electron gas. Hence, the total phonon-drag thermopower may be of either sign and will be smaller in magnitude than either S_g^N or S_g^U.

An interesting departure from (2.36) can arise if S_g^U is dominant at relatively high temperatures and the Fermi surface does not touch the Brillouin zone boundaries, the situation depicted in Fig. 2.7. Although phonons of relatively small wave vector may induce N processes, U processes require phonons of wave vector $\mathbf{q} > \mathbf{q}_{min}$, where $\mathbf{q}_{min}$ is twice the distance of closest approach of the Fermi surface to the nearest zone boundary. Consequently, as the temperature is lowered to the point where the number of thermally excited phonons that have wave vectors in excess of $\mathbf{q}_{min}$ decreases exponentially, S_q^U should also fall off exponentially. Such behavior has been observed in the alkali metals [60M1, 63G1], and from the data the distance of closest approach of the Fermi surface to the zone boundaries has been estimated with results that are in good agreement with band structure calculations.

The low-temperature behavior of the thermopower of polyvalent metals with more complicated Fermi surfaces often exhibits a more complicated pattern as well (Fig. 2.5). An interpretation of such behavior, again invoking phonon drag, allows for at least qualitative understanding of the effects in terms of certain Fermi surface features. For example, if a portion of the Fermi surface is a thin, lens-shaped object (Fig. 2.8), phonons of wave vector approximately equal to the thickness of the lens induce large changes in

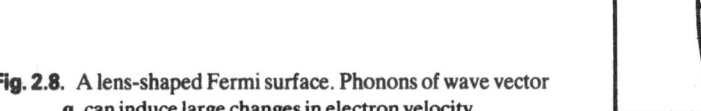

Fig. 2.8. A lens-shaped Fermi surface. Phonons of wave vector $\mathbf{q}_c$ can induce large changes in electron velocity.

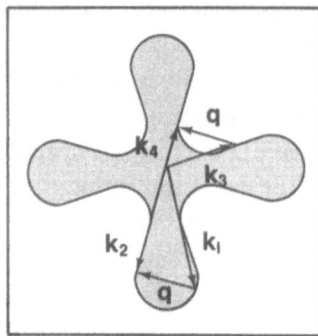

Fig. 2.9. The same phonon, wave vector **q**, can give rise to negative phonon drag (transition $\mathbf{k}_1$ to $\mathbf{k}_2$) or positive phonon drag (transition $\mathbf{k}_3$ to $\mathbf{k}_4$) for this somewhat oddly shaped Fermi surface. A Fermi surface resembling this does exist in tungsten.

electron velocity and thus can make a substantial contribution to phonon drag arising from that part of the Fermi surface. As the temperature is increased from near $T = 0$ K, a peak in S_g is, therefore, likely to develop near that temperature at which phonons of wave vector $\mathbf{q}_c$ are thermally excited in significant numbers. Such arguments appear to account fairly well for the observation on single crystals of magnesium, zinc, and cadmium [70R1].

In the case of a simply connected Fermi surface, which is, however, not spherical, electrons from different portions of the Fermi surface will contribute differently to S_g. Figure 2.9 shows a somewhat pathological Fermi surface where the filled portion is shaded. A phonon of wave vector **q** can induce N transitions, such as $\mathbf{k}_1 \to \mathbf{k}_2$ as well as $\mathbf{k}_3 \to \mathbf{k}_4$. If we remember that **v** is always normal to the Fermi surface, then, in the first of these, $\Delta\mathbf{v}$ is approximately parallel to the wave vector of the absorbed phonon, whereas in the second, $\Delta\mathbf{v}$ is approximately antiparallel to **q**, giving rise to negative and positive contributions to S_g, respectively. Note that in both instances the scattering process is an N event.

As a general rule, applicable to N as well as U processes, electron–phonon scattering will result in a negative phonon-drag thermopower if the **q** vector of the absorbed phonon crosses an occupied (shaded) region of the Brillouin zone; S_g will be positive if the **q** vector crosses an unoccupied region of the Brillouin zone. Evidently, different portions of a nonspherical Fermi surface may contribute with different magnitudes and sign to the total, measured S_g. As we shall see later, such arguments are critical in accounting for changes in S_g due to alloying.

It is well known that the true phonon spectrum of any solid must depart drastically from the dispersionless behavior assumed in the Debye approximation. Dispersion and anisotropy of the phonon spectrum will, of course, exert a profound effect on S_g, which should be apparent at moderate to high temperatures where phonons of large wave vector play a dominant role in electron–phonon scattering. For example, in most metals the decay of S_g with increasing temperature for $T > \theta_D/5$ is far more rapid than $1/T$. This rapid falloff can be understood in light of the true phonon spectrum [64B1].

To summarize, then, we have seen how a nonequilibrium phonon distribution can generate a thermoelectric field through electron–phonon interaction. Elementary arguments predict a phonon-drag thermopower that has qualitatively the correct temperature dependence, but whose magnitude is far in excess of the observed. Consideration of the role of N and U processes, of nonspherical Fermi surfaces, and of phonon dispersion can account for these discrepancies.

2.7 Thermopower of Alloys

Impurities dissolved in a pure metal modify the total thermoelectric power by their influence on both the diffusion and the phonon-drag contributions. Insofar as S_d and S_g may be treated as separate, additive contributions in the pure metal, it is reasonable to treat the changes induced by impurities also independently. However, it is well to bear in mind that recent theoretical work [71N1] has suggested that this additivity may not be valid and that one may have to consider an "interference" term. The calculated magnitude of this interference term is, however, so small that it appears unlikely that it could be identified unambiguously.

2.7a Diffusion Thermopower

We shall start with the general expression for S_d, Eq. (2.25), which we shall rewrite in terms of the resistivity $\rho = 1/\sigma$:

$$S_d = -\frac{\pi^2 k^2 T}{3e}(\partial \ln \rho/\partial \varepsilon)_\eta \tag{2.38}$$

Let us assume the validity of Matthiessen's rule

$$\rho = \rho_i + \rho_j \tag{2.39}$$

where ρ is the total resistivity, ρ_i is the "ideal" resistivity of the pure solvent metal, and ρ_j is the resistivity attributable to scattering of conduction electrons by impurities of type j. To simplify the discussion, we shall consider only one type of impurity; extension to multicomponent alloys is straightforward.

Substitution of (2.39) into (2.38) gives

$$S_d = -\frac{\pi^2 k^2 T}{3e}\left[\frac{\partial \ln (\rho_i + \rho_j)}{\partial \varepsilon}\right]_\eta \qquad (2.40)$$

and a bit of algebra leads to the Friedel relation

$$\frac{\Delta S_d^j}{S_d} = -\frac{1 - x_j/x_i}{1 + \rho_i/\rho_j} \qquad (2.41)$$

where ΔS_d^j is the difference between S_d, the diffusion thermopower of the binary alloy, and S_d^i, the diffusion thermopower of the pure solvent metal; x_i and x_j are given by

$$x_i = -\left(\frac{\partial \ln \rho_i}{\partial \ln \varepsilon}\right)_\eta \qquad x_j = -\left(\frac{\partial \ln \rho_j}{\partial \ln \varepsilon}\right)_\eta \qquad (2.42)$$

Tables 2.1 and 2.2 list experimental values of x_j for various impurities in copper and silver. With two exceptions these parameters are all positive and appear to cluster about a value near unity with magnitudes showing some tendency to increase with increasing impurity valence, indicating, a not altogether unexpected result, that as the residual resistivity, i.e., the scattering, increases, so does the dependence of the scattering cross section on electron energy. The two notable exceptions, CuCd and AgMg, appear to be truly anomalous since similar anomalous behavior is manifested in the pressure dependence of the resistivity (see below).

An alternative, equivalent, but more convenient form is the Gorter–Nordheim relation [35N1]

$$S_d = S_d^j + \frac{\rho_i}{\rho}(S_d^i - S_d^j) \qquad (2.43)$$

where S_d^j, known as the characteristic thermopower of the impurity of type j in the particular solvent metal, is related to the Friedel parameter x_s by

$$S_d^j = \frac{\pi^2 k^2 T}{3e\eta} x_j \qquad (2.44)$$

From the definitions of x_j and S_d^j it is apparent that these parameters are independent of impurity concentration, at least within the restrictive assumptions of the theory, i.e., the validity of Matthiessen's rule (which is generally violated in real systems [72B1]) and the presumption that alloying does not alter the Fermi surface and thereby change S_d^i [67B2]. However, scattering anisotropy, in particular, a major cause for deviations from Matthiessen's rule, can similarly lead to significant departures from the Gorter–Nordheim relation [72G4]. Guénault [74G5] has shown that the widely divergent results for S_d^j in homovalent copper alloys deduced from high-temperature and low-temperature measurements can be understood from such considerations.

If Eq. (2.43) is taken as a reasonable first approximation, it then follows that a plot of S_d vs. $1/\rho$ at constant temperature should yield a straight line whose intercept is S_d^j and whose slope is $\rho_i(S_d^i - S_d^j)$. It has become common practice to present results on the thermopowers of alloys in terms of such Gorter–Nordheim plots, such as shown in Fig. 2.10. However, a slightly modified representation of experimental data may be preferable [65S1].

Using Eq. (2.39) we can rewrite (2.43) as

$$S_d = S_d^i + \frac{\rho_j}{\rho}(S_d^i - S_d^j) \tag{2.45}$$

Since both S_d^i and S_d^j should be linear functions of T, a plot of S_d/T as a function of $1/\rho$ at constant impurity concentration should again give a straight line whose intercept is S_d^i. Plotting results in this manner will reveal whether the ideal diffusion thermopower S_d^i is indeed unaffected by alloying, a condition that is not likely to prevail if alloying alters the Fermi surface. Such a set of straight lines obtained for a group of copper–nickel alloys is shown in Fig. 2.11, and it is apparent that as nickel is added to copper the diffusion thermopower associated with phonon scattering S_d^i shows marked changes. These changes may be due to reduction in the size of the "necks" of the Fermi surface with increasing nickel concentration as the conduction electron concentration is diminished.

An expression similar to Eq. (2.43) but of more general validity was derived by Kohler [49K1] using the variational solution of the transport equation, namely,

$$S_d = \frac{W_i}{W} S_d^i + \frac{W_j}{W} S_d^j \tag{2.46}$$

where W, W_i, and W_j are the total, ideal, and impurity electronic thermal resistivities, respectively. Equation (2.46) reduces to (2.43) if the Wiedemann–Franz law holds, that is, if $\rho_0 = L_0 T W_0$, where L_0 is the Lorenz number $\pi^2 k^2 / 3e^2$.

Table 2.1. Volume Coefficient of Residual Resistivity and Impurity Thermopower Parameter for Copper Alloys

Impurity	Temperature					
	20 K		273 K		300 K	
	α_j	x_j	α_j	x_j	α_j	x_j
Zn	0.30^c		0.08^b	0.63^h	0.43 ± 0.11^a	0.59^h
				0.89^k	0.41^d	
Cd	-0.30^c			-0.58 ± 0.29^k		
Al			0.54^b	0.49^f	0.50^a	
				0.26^g		
Ga	0.90^c		1.11^b	0.26^h	1.00 ± 0.04^a	0.28^h
			1.53^e	0.20^g	1.10^d	0.14^m
In	0.70^c			0.63^k		
Si			1.03^b	1.04^g	1.04 ± 0.04^a	0.63^m
Ge	0.60^c		0.76^b	0.71^h	0.97 ± 0.03^a	0.74^h
			1.11^e	0.78^g	0.96^d	0.72^m
				0.69^k		
				0.56^m		
Sn	0.90^c		0.97^b	0.99^g		
				1.26^m		
				1.30^k		
As			0.58^b	0.84^h	0.75^d	0.90^h
Sb			0.90^b	1.55^g		1.54^m
				1.73^k		

^{a} 49B1. ^{f} 67W1.
^{b} 69L1. ^{g} 64K1 and 70F4.
^{c} 65D2. ^{h} 65C1.
^{d} 74H1. ^{k} 67S3.
^{e} C. L. Foiles, unpublished data. ^{m} 58D1.

Calculation of the pertinent parameter, x_j or S_d^j, is by no means a straightforward matter. If, following Mott's early suggestion, an impurity of valence Z_j in a host metal of valence Z_i is represented as a point charge $(Z_j - Z_i)e$ subject to Fermi–Thomas screening by the conduction electrons, it is possible to derive an analytic expression for x_j [53F1]. That expression, which depends only on the Fermi wave vector and the Fermi–Thomas

Table 2.2. Volume Coefficient of Residual Resistivity and Impurity Thermopower Parameter for Silver Alloys

Impurity	Temperature							
	20 K		100–250 K		273 K		≥300 K	
	α_j	x_j	α_j	x_j	α_j	x_j	α_j	x_j
Zn	0.90^c			0.52^n	0.67^b	0.57^g 0.56^h	2.33^a	0.56^h
Cd	1.00^c			0.41^n	1.11^b	0.21^g 0.27^p	2.92^a	
Ga					1.05^b	0.79^h 0.84^g		0.84^h
Al						0.59^g		
In	1.60^c	0.96^i		0.93^n	1.43^b	0.84^g	1.50^a	
Tl	1.40^c	0.90^i		1.24^n	1.33^b			
Ge	0.80^c	0.56^i		0.88^n		1.03^f 0.90^g 0.92^h		0.98^h
Sn					1.04^b	1.25^g		
As					0.36^b	1.10^g 0.79^h		0.84^h
Pb					0.84^b			
Sb	0.90^c					1.35^n		1.38^g
Mg					-1.82^b	-0.59^g		

[a] 49B1. Based on measurements on a single alloy.
[b] 69L1.
[c] 65D2.
[f] 67W1.
[g] 64K1 and 70F4.
[h] 65C1.
[i] 67G1.
[n] 68W1.
[p] P. A. Schroeder, unpublished data.

screening radius, is based on the assumption that electron–impurity scattering may be calculated in the Born approximation. The result suggests that all impurities in a given solvent metal will have the same characteristic diffusion thermopower. This is clearly not the case, although the observed values of S_d^j are often of roughly the same magnitude and sign.

It has, of course, been recognized for many years that the Born approximation is not valid for scattering of conduction electrons by nonhomovalent impurities in metals. Calculations of the residual resistivity using a phase-shift procedure, wherein the scattering potential is adjusted so as to satisfy the Friedel condition, give results in quite good agreement with experiment [60Z1]. Using these same methods, one can also calculate x_j, and one does find that this parameter now depends on the impurity potential

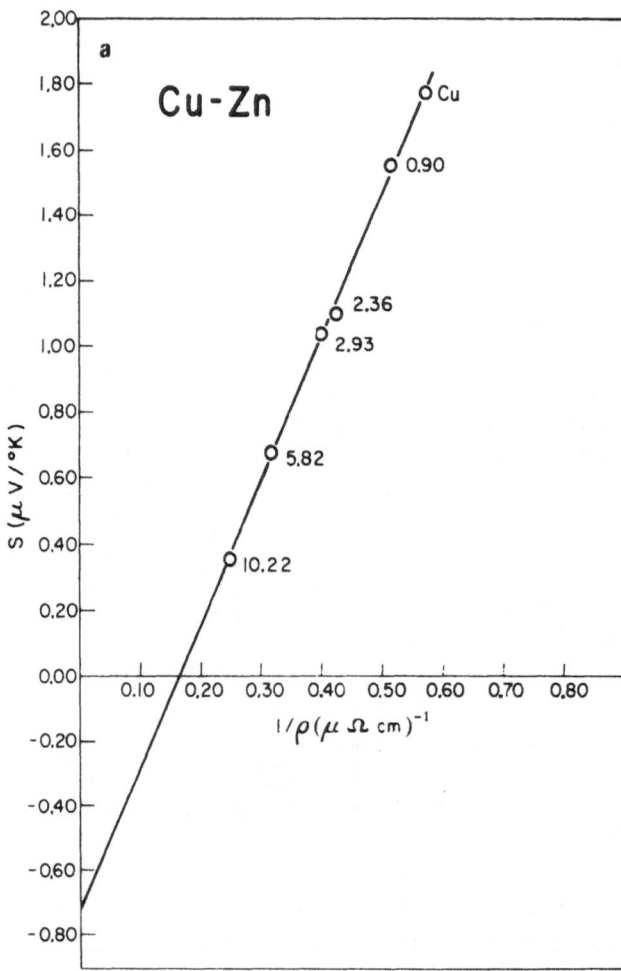

Fig. 2.10. A Gorter–Nordheim plot for (a) copper–zinc alloys at 300 K (from [63H1]), and (b) AgIn, AgGe, and AgTl alloys at various temperatures (from [68W1]).

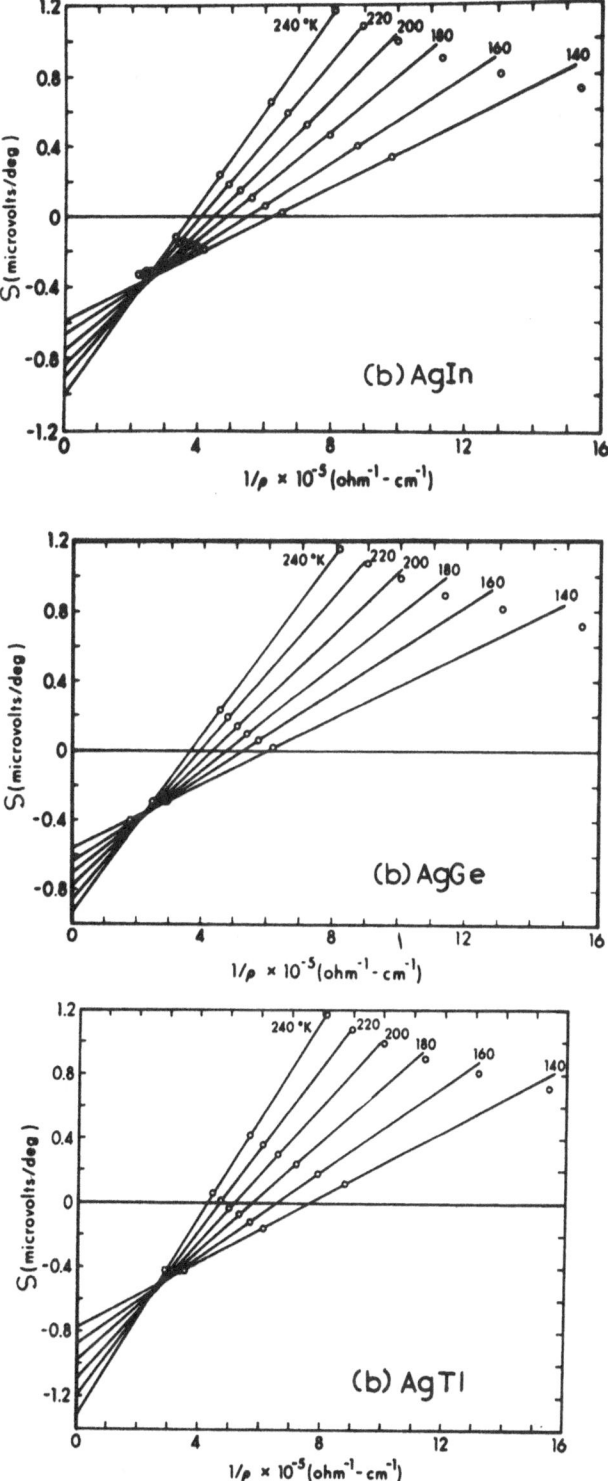

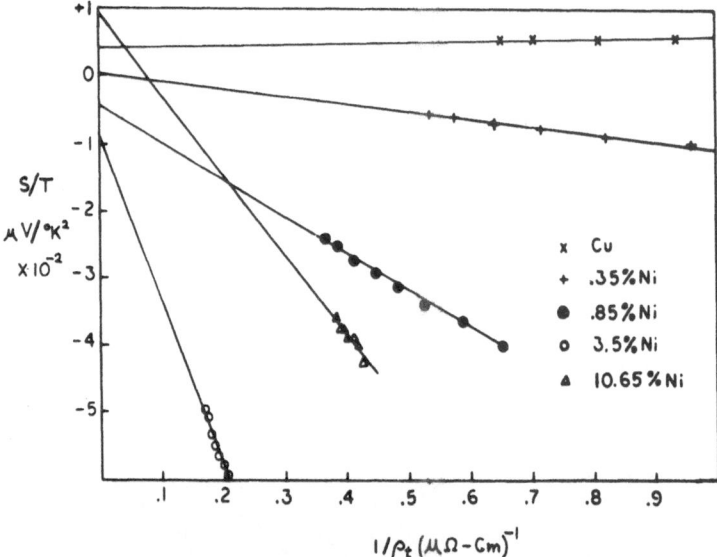

Fig. 2.11. A modified Gorter–Nordheim plot for Cu–Ni alloys. Here S_d/T is plotted against $1/\rho$ as T is varied and the residual resistance is kept fixed (from [65S1]).

as well as the properties of the host metal [56D1]. However, whereas the calculated residual resistivities are relatively insensitive to the details of the scattering potential provided the proper Friedel condition is satisfied, the same is definitely not true for x_j. In calculations of this parameter different assumptions concerning the pseudopotential yield substantially different results [73F1], and, to some extent, data on thermopowers of alloys could assist in selecting the "best" from a variety of possible pseudopotentials. Two alloy systems are of particular interest in this connection.

Because of their nearly spherical energy surfaces, the alkali metals and their alloys are most advantageous objects for study. Moreover, for these metals a local pseudopotential approximation should be quite good; the same is not true for the noble metals. Unfortunately, solid solubility in the alkalis is extremely limited, and so relatively few alloy systems can be studied. Potassium is a reasonably good host, accepting up to 2.5 at.% sodium and forming continuous solid solutions with rubidium and cesium. This system has been studied by Guénault and MacDonald [61G1, 63G1] with very interesting results.

We have already seen that elementary arguments predict that in the impurity scattering region, where $\tau \propto \varepsilon^{-1/2}$ (constant mean free path), S

should be negative. It is, thus, somewhat surprising to find that in the more concentrated K–Rb and K–Cs alloys S is positive, suggesting that τ decreases much more rapidly with energy than $\varepsilon^{-1/2}$. Using the phase shifts calculated by Meyer *et al.* [67M1], Thornton *et al.* [68T1] calculated the parameters x_j for various alkali–alkali alloys with results which, though differing significantly in numerical value from experiment, do predict the observed anomalous signs. As shown by Srivastava and Sharma [69S1], calculated x_i values are sensitive to the choice of pseudopotential, and the same is undoubtedly also true for x_j of the alkali–alkali system. Somewhat different choices of these potentials might well lead to better numerical agreement, but the work of Thornton clearly shows that current theoretical techniques are now adequate to account for the experimental observations at least qualitatively.

The second alloy system of interest is that of the noble metals. Transport properties of alloys of copper, silver, and gold with numerous solute metals have been studied extensively in the past. These metals have fairly wide solubility ranges, are readily available, are nonreactive, and have advantageous metallurgical properties. Resistivities and thermopowers of the "pure" metals and of numerous alloys have been measured [65M1, 72B2], and the x_j values are shown in Tables 2.1 and 2.2.

As Friedel [56F] pointed out, one should expect to find a close correlation between the thermoelectric parameter x_j and the volume dependence of the impurity resistivity ρ_j. The relationship follows directly from the free-electron model, if it is assumed that the dominant effect of a reduction in atomic volume, brought about by application of hydrostatic pressure, is an increase in Fermi energy. Recalling that $\varepsilon_F \propto n_0^{2/3} \propto V^{-2/3}$, we obtain

$$\alpha_j \equiv \frac{d \ln \rho_j}{d \ln V} = \frac{d \ln \rho_j}{d \ln \varepsilon} \frac{d \ln \varepsilon}{d \ln V} = \tfrac{2}{3} x_j \tag{2.47}$$

Tables 2.1 and 2.2 also list the values of α_j as deduced from measurements at various temperatures. There are, obviously, substantial discrepancies between results of different investigators, rather greater than for the x_j values. Part of the difficulty is undoubtedly related to the problem of extracting the pressure dependence of the impurity resistivity from that of the total resistivity, especially at the higher temperatures [61D1].

In any event, a glance at these tables reveals that (except for AgMg, AgCd, and CuGa) the ratio α_j / x_j takes on values that range between about 0.5 and 1.3, i.e., which are at least of the correct magnitude. It is also noteworthy that in the two alloy systems, AgMg and CuCd, for which x_j is negative, α_j also exhibits an anomalous sign change. With regard to calculations of these parameters, there has been some modest success [70D1], although the use of the Born approximation in these calculations has cast doubt on some of the results [73F1].

Returning briefly to the matter of pure metals, it is clear that a relationship akin to Eq. (2.47) should exist also between x_i and $\alpha_i = d \ln \rho_i / d \ln V$. Here, however, the situation is complicated by the fact that ρ_i, unlike ρ_j, depends even in the free-electron approximation on the dynamic properties of the lattice and the strength of the electron–phonon coupling as well as the Fermi energy. If the phonon spectrum is treated in the Debye approximation and U processes are neglected, ρ_i is given by the Bloch–Grüneisen expression [68B1], which at high temperature reduces to

$$\rho_i = ACT/\theta_D^2 \tag{2.48}$$

where A involves various physical constants as well as parameters characterizing the electron system, C is a measure of the strength of the electron–phonon coupling, and θ_D is the Debye temperature. Thus, one obtains

$$\alpha_i = -\tfrac{2}{3}x_i + 2\gamma + \partial \ln C / \partial \ln V \tag{2.49}$$

where γ is the Grüneisen constant

$$\gamma = -\partial \ln \theta_D / \partial \ln V$$

Table 2.3 shows the values of α_i and $2\gamma - \tfrac{2}{3}x_i$ for lithium, sodium, and potassium. The correlation is quite good, even neglecting the last term of Eq. (2.49); Barnard [71B1] has argued that this term is indeed small.

Extension of (2.43) to multicomponent alloys gives the general expression

$$S_d = \frac{\rho_i}{\rho}S_d^i + \sum_j \frac{\rho_j}{\rho}S_d^j \tag{2.50}$$

Table 2.3. Values of α_i and $2\gamma - \tfrac{2}{3}x_i$ for Lithium, Sodium, and Potassium

Metal	α_i	$2\gamma - \tfrac{2}{3}x_i$
Li	−0.8	−2.66
Na	4.43	4.35
K	5.29	5.34

In a few studies of the thermopower of ternary alloys, the application of Eq. (2.50), using values for S_d^j determined from work on binary alloys, has given good agreement with observation [67B3].

2.7b *Phonon Drag*

In the above discussion we have ignored the phonon-drag term. Obviously, comparison of experimental data with the Gorter–Nordheim relation requires that the role of S_g be properly recognized. In practice, this means that data are frequently obtained either at temperatures so high that phonon drag is unimportant already in the solvent metal, or at temperatures so low that an approximate separation of the total thermopower into diffusion and phonon-drag contributions can be achieved by plotting S/T vs. T^2 [see Eq. (2.36)]. Prior to about 1960, that is, before the importance of phonon drag was widely appreciated, attempts to interpret thermopower data for alloys within the framework of the Friedel or Gorter–Nordheim formalism often gave inconsistent and confusing results, especially near liquid nitrogen temperature, where S_g is usually quite large. These difficulties are, however, largely removed once the diffusion thermopowers of the pure metal and of the alloys are abstracted from the total measured thermopowers. However, impurities also have a profound effect on the phonon-drag thermopower. We shall defer discussion of these influences to Chapter 4, which is devoted to the detailed consideration of phonon drag.

3 | Techniques in Thermoelectric Measurements

3.1 Introduction

We shall digress here from the theoretical discussion of the TEP to consider in some detail the experimental techniques that have been employed in the past and that are currently used to measure thermoelectric effects. These measurements fall naturally into four categories: the first three are concerned with the determination of conventional thermoelectric parameters, the TEP, S, the Peltier heat (measurement of Π), and the Thomson coefficient μ; the last is concerned with other parameters, specifically Π_e, or $G = 1/\Pi_e$, introduced at the end of Chapter 1.

The various thermoelectric parameters are, of course, interrelated through the Kelvin relations, Eq. (1.8), and, in the case of G, through the Wiedemann–Franz ratio. That is,

$$\frac{S}{GT} = \frac{\Pi_e}{\sigma T} = L \tag{3.1}$$

Consequently, in terms of obtaining fundamental information concerning relaxation times and/or band structure, there is little to recommend one effect in preference over another. However, as we shall see, experimental considerations often favor the measurement of a particular effect in a certain temperature regime.

Generally, the thermopower S is the simplest to measure, although at very low temperatures it may be more convenient to measure Peltier heats.

As was pointed out in Chapter 1, measurement of either S or Π yields the difference between these quantities for two dissimilar conductors. To obtain the absolute TEP or Peltier coefficient, a thermocouple or Peltier junction of which one arm is one of the standard metals should be employed.

3.2   Seebeck Effect

There are essentially two techniques for measuring S_{AB}, which we denote as the integral and the differential methods. Both are based on Eq. (1.4)

$$\Delta V = V_b - V_a = \int_{T_1}^{T_2} (S_B - S_A)\, dT \tag{1.4}$$

and its derivative form

$$\frac{d(\Delta V)}{dT_2} = S_{AB}(T_2) = S_B(T_2) - S_A(T_2) \tag{3.2}$$

In the integral method the temperature T_1 of one junction of the thermocouple is maintained fixed, and ΔV is measured as a function of T_2. Differentiation then gives $S_{AB}(T_2)$, from which $S_B(T_2)$ can be deduced if metal A is a standard, such as lead or platinum. Differentiation is conveniently accomplished either graphically, i.e., by plotting ΔV vs. T_2 and measuring the slope at the desired values of T_2, or numerically with a small computer.

In the differential or incremental method a small temperature difference $\Delta T = T_2 - T_1$ is maintained between the two junctions, and the average temperature is varied. In that case we have

$$S_{AB} = \Delta V / \Delta T \tag{3.3}$$

When this technique is used, it is imperative that ΔT be sufficiently small so that neither S_B nor S_A changes substantially in that temperature interval; at the same time, ΔT must be large enough to generate a voltage that can be measured to the desired precision.

The integral method allows for rather simpler apparatus and electronics because T_1 can be the temperature of the cryogenic bath, such as liquid nitrogen or helium, into which one junction is submerged, and need not be measured as it must be in the incremental method. On the other hand, with the integral method large temperature gradients are required if measurements are performed between 4.2 and 80 K or 77.7 and 300 K. For this reason, the method is best suited when the samples are in the form of long wires. Since wires are generally polycrystalline, the measurement gives some average of the thermopower tensor unless the sample is cubic, in which case the thermopower tensor is a multiple of the unit tensor.

To obtain the coefficients of the thermopower tensor, it is necessary to use single-crystal specimens, and because these are usually of limited length, they do not lend themselves well to the integral method. Similarly, any materials which cannot be extruded or drawn into wires are best tackled using the incremental method.

In all thermopower measurements one must exercise care to eliminate spurious thermal emfs which arise whenever leads that are not strictly homogeneous pass through a temperature gradient. Here we include homogeneity in purity as well as vacancy and dislocation content. In our experience, the best leads are pure Pb wires, but as these are quite fragile, Cu leads joined to the thermocouple sample in the cryogenic fluid are often satisfactory.

As an example of the apparatus used in the integral method, we show in Fig. 3.1 the one used by Henry and Schroeder [63H1], which, in somewhat modified form, has been used at Michigan State University for a number of years. The cold junction of the Pb-sample thermocouple is normally immersed in liquid helium (at 4.2 K or pumped to about 1.1 K) or in liquid nitrogen. During an experimental run the refrigerant level is partway up the brass rod. The hot junction of the thermocouple is soldered with low-melting-point solder to a copper block in which suitable thermometers are embedded. For the apparatus shown in the diagram, for example, the platinum resistance thermometer is used for the region from 12 to 300 K and the carbon thermometer below 12 K. In recent years we have turned to germanium resistance thermometers for the range from 1.5 to 100 K. Space may be provided in the copper block for a sample for resistivity measurements. The copper block is surrounded by a copper cylinder to provide a nearly isothermal enclosure for the block. The cylinder, in turn, is covered with asbestos or some other thermal insulator to reduce temperature fluctuations due to convection currents in the refrigerant vapor. The entire assembly is maintained at a given temperature by means of a heater attached to the base of the cylinder. Quite good results can be obtained by manual control of the heater to maintain a constant temperature T_2 of the hot junction while measuring the thermoelectric voltage of the thermocouple

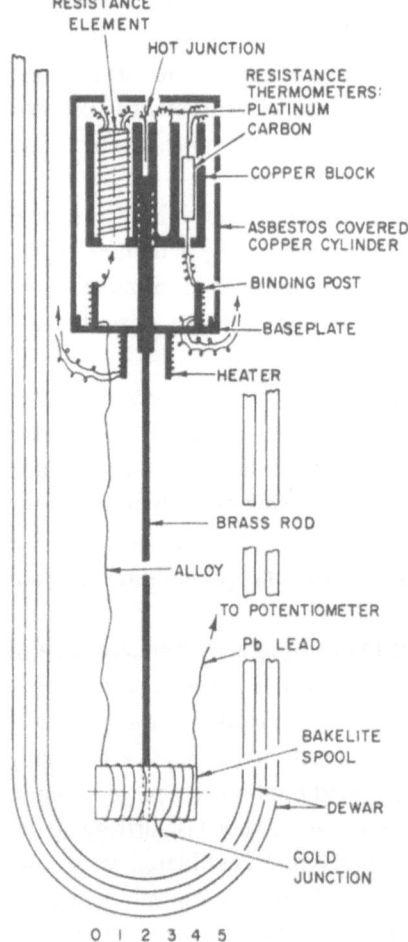

RESISTANCE
ELEMENT

HOT JUNCTION

RESISTANCE
THERMOMETERS:
PLATINUM
CARBON

COPPER BLOCK

ASBESTOS COVERED
COPPER CYLINDER

BINDING POST

BASEPLATE

HEATER

BRASS ROD

ALLOY

TO POTENTIOMETER

Pb LEAD

BAKELITE
SPOOL

DEWAR

COLD
JUNCTION

0 1 2 3 4 5
    cm

Fig. 3.1. Cryostat for the measurement of thermopower using the integral method (from [63H1]).

and the appropriate thermometer. It is best, however, to control the power input to the heater by means of a thermocouple or resistance sensor to maintain constant temperature. Such an arrangement permits a more leisurely experiment and allows the nearly simultaneous measurement of different samples by means of external switching between samples.

The advantages of the method are; (1) Simplicity—no vacuum system is required except to pump on liquid helium if the cold-junction temperature is to be reduced below 4.2 K. (2) Wide temperature range. (3) Good accuracy—measurements of S to ± 0.015 μV/K are readily attained and are generally sufficient for most purposes. An outline of the computer analysis of the raw data can be found in the original paper. A disadvantage is that sharp features of the TEP tend to get smeared out unless great care is taken to gather a sufficient density of data points in the region of the feature and, thereafter, to maintain and use this density of points during the numerical differentiation process. In the above method it is assumed that any spurious emfs remain constant or change so slowly that their effect on the thermopower measurement is minimal.

A simple cryostat designed for the incremental method which operates over the same wide temperature range is illustrated in Fig. 3.2 [70R1]. The thermal emf of a Pb–sample thermocouple is measured while a small temperature difference is maintained along the sample. Lead wires are attached to the sample at points C and H, which are also the locations of the cold and hot junctions of a Au–Fe vs. chromel thermocouple used to determine the small temperature difference ΔT of Eq. (3.3). A second

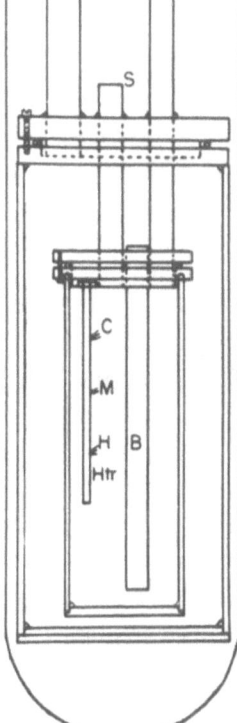

Fig. 3.2. Cryostat for the measurement of thermopower using the incremental method. C and H refer to cold and hot junctions of gold–iron vs. chromel thermocouples attached to the crystal at the same points as the Pb–crystal thermocouple junction. Another gold–iron vs. chromel thermocouple is attached at M. Htr is a heater wound on the crystal. B is a binding post and S is an epoxy seal through which the thermocouple leads were extracted (from [70R1]).

thermocouple, whose cold junction is immersed in the refrigerant and hot junction attached at point M, monitors the sample temperature. A heater wrapped on the inner can controls its temperature.

For measurements below 4.2 K the outer can is removed and the temperature of the inner can and sample is adjusted by pumping on the cryogenic bath. Between 4.2 and about 35 K the space between the inner and outer cans is filled with He gas at a controlled pressure, to provide a large heat leak to the bath in the temperature region where a large heat flow is required to produce a measurable temperature gradient in the crystal. Above about 35 K the space between the cans is evacuated. Throughout the experiment the sample temperature is maintained by electronic control of the heater current, using the thermocouple attached at M as sensor.

Continuous Pb leads are used, passing up the pumping tube P_1 and out of the cryostat through epoxy seals, so as to minimize spurious thermals. Such precautions are not necessary for the leads connecting to the measuring thermocouples since their emfs are generally much greater than those from the sample. In any event, the effect of spurious thermals is overcome by taking data for a series of temperature differences ΔT, ranging from 0 to 0.5 K near 300 K and from 0 to 0.1 K near 4.2 K, and then plotting ΔT vs. ΔT. The slope of the resulting straight line equals the thermopower of the sample thermocouple and the spurious thermals appear only as nonzero intercepts on the ΔV axis.

The same apparatus can also be used for measurements of thermal conductivity, in which case the inner can is always evacuated. A modification of the method which permits rather faster collection and analysis of data has been developed by Caskey et al. [69C1].

3.3 Peltier Effect

Because the determination of Peltier heats requires sensitive calorimetry, they have rarely been measured in the past. However, as Trodahl [69T1] has pointed out, there may be considerable advantage in measuring Peltier heats at temperatures below 4.2 K, because in this region thermopowers are generally small, requiring expensive and sophisticated instrumentation for the measurement of small voltages, whereas carbon resistors have a very high sensitivity, permitting precise relative calorimetry. The measurement of Peltier heats, using Trodahl's method, has the added advantages that a magnetic field may be applied and that the sample

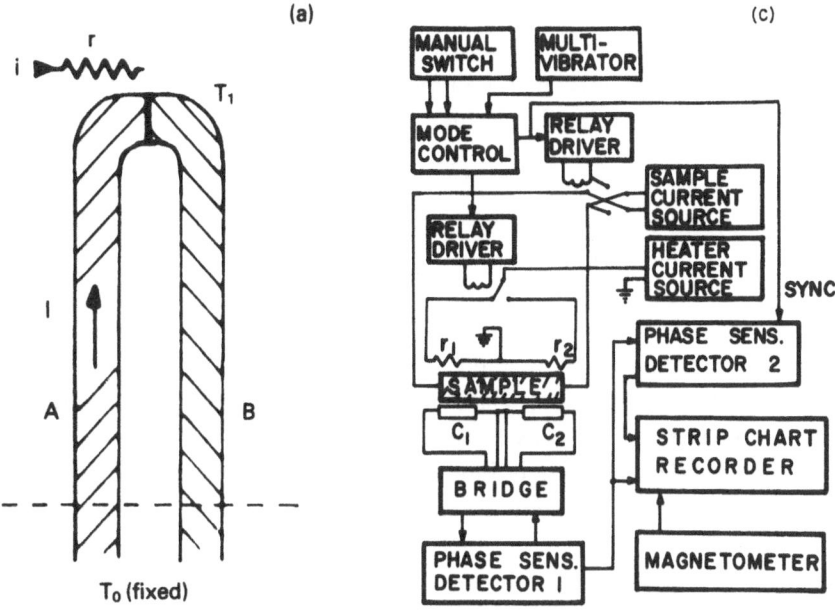

(a)

(c)

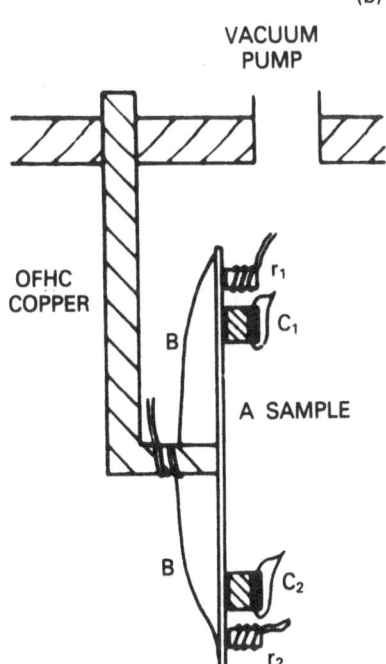

(b)

Fig. 3.3. Peltier measurement at low temperatures. (a) Schematic diagram of Peltier junction. (b) Diagram of double junction. The center of the sample was soldered to the copper arm. The copper blocks supporting the carbon resistors C_1, C_2 and calibrating heaters r_1, r_2 were attached to the polycrystalline samples with a crimp contact. In the measurements on single crystals the thermometers C_1, C_2 were glued directly to the sample with GE7031 varnish, and the carbon blocks on which r_1 and r_2 were mounted were soldered directly to the junction. (c) Block diagram of electronic circuitry for Peltier measurements (from [69T1]).

resistance need not be small, a condition required if superconducting measuring devices are employed (see below).

The apparatus is relatively simple and is illustrated in Fig. 3.3. Referring to Fig. 3.3a, we see that Peltier heat

$$\dot{Q}_p = \Pi_{AB}I \tag{3.4}$$

is produced as current I passes through the junction of metals A and B, resulting in a temperature difference

$$(T_1 - T_0)_P = \Delta T_P = W\dot{Q}_P \tag{3.5}$$

where W is the thermal resistance between the junction at temperature T_1 and the fixed temperature T_0. The temperature of the junction will, of course, also be influenced by Joule heating, but this effect can be accounted for by reversal of the current. Finally, one must be cognizant that Thomson heat will also be generated in the two arms, but since this effect depends on the temperature difference, it can be made negligibly small if $\Delta T = \Delta T_J + \Delta T_P$ is kept sufficiently small. Calibration of the system is accomplished by means of a resistor r attached to the junction. Following the determination of ΔT_P, the current through the junction is set to zero, and the current i through the known resistor which will result in the same temperature difference is determined. The Peltier coefficient is then given by

$$\Pi_{AB} = i^2 r/I \tag{3.6}$$

The smallest Peltier coefficient that can be measured to 1% accuracy is given by [69T1]

$$\Pi_m = \Delta T_m (12.59 L_A)^{1/2} \times 100 \tag{3.7}$$

where ΔT_m is the smallest detectable temperature change, and L_A is the Lorenz number of the sample. Metal B is assumed to be a superconductor. For $\Delta T_m \simeq 10^{-7}$ K at 2 K, $\Pi_m \simeq 6 \times 10^{-9}$ V, and the sensitivity is approximately doubled by using the double-junction technique described below, and is equivalent to the measurement of a TEP of 1.5×10^{-9} V/K at 2 K.

The temperature dependence of the sensitivity is largely determined by the temperature dependence of ΔT_m. Neither electrical nor thermal conductivities appear directly in Eqs. (3.6) or (3.7) (although their ratio is implicitly contained in L_A) nor do the dimensions of the sample; consequently, the method may be particularly advantageous for alloy measurements. (Many of the methods outlined in Section 3.6 for measuring small voltages depend on the sample resistance being very small.)

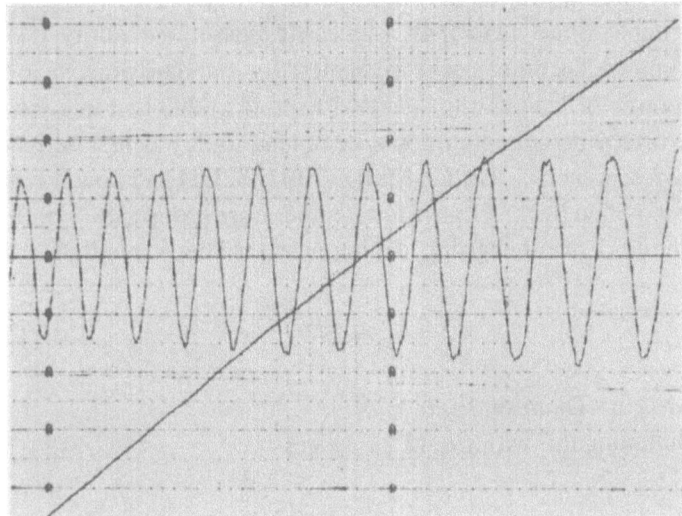

Fig. 3.4. Strip chart recorder trace of Peltier oscillations in tin using arrangement of Figure 3.3b and c. The magnetic field is varied from about 14 kG at left to 16 kG at right. The output trace, proportional to Π, was obtained from PSD2 of Figure 3.3c (from [69T1]).

The double-junction apparatus is illustrated in Fig. 3.3b. Here r_1 and r_2 are calibrating resistors, c_1 and c_2 are carbon resistance thermometers, A is the sample, and B a superconductor completing the circuit. The two carbon resistors constitute two arms of an ac Wheatstone bridge which is driven and whose imbalance is detected by a phase-sensitive detector (PSD). In some of his measurements, Trodahl made use of a second PSD tuned to the frequency at which the current through the junctions was reversed. The output of the first PSD was fed into the second, so that the output of the latter was a continuous function of time. This technique was employed in studying the quantum oscillations of the Peltier coefficient of tin [69T1], an example of which is shown in Fig. 3.4.

3.4 Thomson Effect: The Absolute Thermopower of Lead

We have already stated that the thermopower of a thermocouple S_{AB} is

$$S_{AB} = \frac{d(\Delta V)}{dT} = (S_B - S_A) \tag{3.2}$$

where S_A and S_B are the absolute thermopowers of the two arms of the thermocouple. Since $d(\Delta V)/dT$ is an easily measured quantity, if either S_A or S_B is known, the other can be calculated. It is standard practice to take the thermopower of lead as the standard from 0 to 293 K. Here we wish to indicate briefly how the thermopower of lead itself was first derived about 1930 by researchers at Leiden [28B1, 30B1, 32B1] and how it was subsequently modified by the Ottawa low-temperature group [58C1] in 1958. To follow the work of either group, it is important to keep in mind the relation

$$\mu = T\frac{dS}{dT} \tag{1.8a}$$

where μ is the Thomson heat.

Combining this with Eq. (3.2), we get

$$\mu_A = \mu_B + T\frac{dS_{AB}}{dT} \tag{3.8}$$

To obtain S_{Pb} as a function of temperature, it is necessary to find μ_{Pb} as a function of temperature and then perform the integral

$$S(T) = \int_0^T (\mu/T')\, dT' \tag{1.9}$$

The steps followed by the Leiden group were as follows:

1. For tin and lead below the superconducting critical point of tin (3.7 K) $\mu/T = 0$.
2. Between 3.72 and 7.2 K, $\mu/T = 0$ for lead. μ/T for tin can be derived from the thermopower of a Pb–Sn thermocouple using Eq. (3.8).
3. For $20\,K < T < 300\,K$, μ/T is measured directly for a particular alloy "silver-normal" -0.37 at.% gold in silver. Values of μ/T for Pb and Sn are obtained from the thermopower of the corresponding thermocouples, using Eq. (3.8) again. The lower temperature limit approximates the boiling point of liquid hydrogen, which was the refrigerant most readily available in the Leiden laboratories.
4. Between 7.2 and 20 K an interpolation of μ/T for tin was made. Values of μ/T for Pb and silver-normal were obtained by the appropriate thermopower measurement using Eq. (3.8). The interpolation was performed on tin because of the smooth regions below

7.2 K and above 20 K, which made for comparatively straightforward interpolation.

5. At this stage μ/T for Pb is known from 0 to 300 K, and S is obtained
from Eq. (1.9).

The Ottawa group largely replaced the interpolation in step 4 by
measuring the thermopower of a Pb–Nb$_3$Sn thermocouple. The Nb$_3$Sn is
superconducting and, therefore, has zero thermopower up to 17.92 K. The
thermopower of lead up to this temperature is just the measured thermopower of the thermocouple. From Eq. (1.8a) μ for Pb is obtained over
this range. Then only a small interpolation of μ/T for Pb from 18 to 20 K is
required. Thereafter, S is obtained over the full range 0 to 300 K from Eq.
(1.9) as before. It should be noted that the absolute thermopower of Pb as
listed by Christian *et al.*, which is now universally used as a standard,
incorporates two separate measurements of the thermopower of Pb with
respect to silver-normal, which in themselves differ from each other by up to
6%. Up to 100 K, Christian *et al.* use the 1931 results of Borelius *et al.*
[31B1], and above 100 K they use their 1930 results [30B1].

The significant point at the moment is that from 20 to 300 K, *all
present-day absolute thermopower measurements are completely dependent
on the Thomson heat measurements of Borelius and his co-workers of nearly
half a century ago.* Because of its historic interest, its unique position in
thermoelectric research, and because it is found in a journal which is
relatively inaccessible, we consider it worthwhile to describe their research
in some detail.

The principle of the measurement is as follows. The two ends of a wire
are maintained at the same temperature. A current passes through the wire
heating the central region and creating a temperature gradient in opposite
directions on either side of the center. Because the current flows through
these two different temperature gradients, Thomson heats of opposite sign
will be generated on either side of the center. The Thomson coefficient is
calculated from

$$\mu = \frac{\kappa}{I} \cdot \frac{\Delta T_T}{\Delta T_J} \cdot \frac{3A}{y} \Phi \tag{3.9}$$

where κ is the thermal conductivity of the specimen; I is the current; ΔT_T is
the change in temperature produced by the Thomson effect and ΔT_J is the
change in temperature produced by Joule heating, both temperature
changes being measured a distance $\pm y$ from the center; A is the cross-

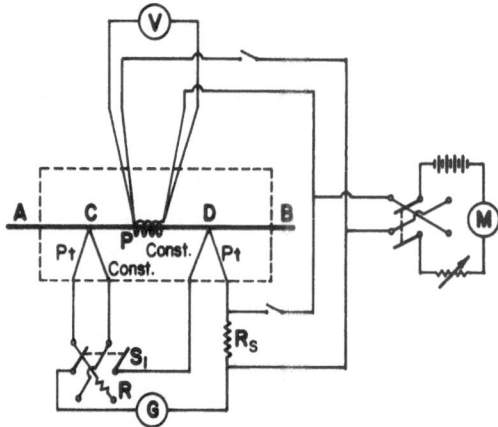

Fig. 3.5. Experimental arrangement used by Borelius and collaborators to measure Thomson coefficients. The dashed line indicates the isothermal enclosure.

sectional area of the wire; and Φ is a correction factor introduced to account for the resistance of and the radiation from the wire changing with the temperature. In the worst cases Φ deviated from 1 by -2.5% and 1.8%. The circuitry is shown schematically in Fig. 3.5.

The mean temperature of the sample wire AB was measured by a platinum resistance thermometer P (consisting of about 20 turns of 0.05-mm-diameter "physically pure" platinum), wound round its midpoint. The thermometer was calibrated in the apparatus at a number of temperatures—melting ice, liquid hydrogen, and nitrogen or oxygen boiling at different pressures. The sample wire itself was covered with a thin layer of enamel lacquer, which enabled the resistance thermometer to be wound directly on the wire and also permitted the direct attachment of the junctions of two platinum–constantan thermocouples to the wire by shellac.

To measure ΔT_J, the reversing switch was positioned to place the thermocouples in series. The cold junctions corresponded to the temperature of the isothermal enclosure shown schematically in Fig. 3.5. The current I through the sample was then turned on and off. When the current flowed, there was Joule and Thomson heating resulting in temperature changes at C and D. Since the Thomson heat was of opposite sign at C and D, its effect canceled for the series connection, and the emf measured by the sensitive galvanometer G therefore corresponded to ΔT_J.

To measure ΔT_T, the thermocouples are connected in series opposition. The emfs caused by ΔT_J should then cancel, and the emf corresponding to ΔT_T is measured. The sensitivity is doubled by reversing the current. Resistance R served to reduce the sensitivity of the galvanometer when the thermocouples were in series aiding. The galvanometer sensitivity could be checked by passing a current, measured by the milliammeter M, through the standard resistance R_s. No details are given of the platinum resistance thermometer measurements. However, the method is fairly obvious from the circuit.

The sample was mounted on a 3-kg block of copper suspended in a vacuum and cooled by direct contact with liquid refrigerant (presumably liquid nitrogen, oxygen, or hydrogen). Its heat capacity was sufficient that, for a liquid hydrogen run, after the hydrogen had evaporated, its temperature rose slowly enough to permit measurements during the temperature rise. The various wires were thermally tied to the block with the ubiquitous cigarette paper used for good electrical insulation.

Equation (3.9) contains the thermal conductivity κ, which was measured by independent experiments. Equation (3.9) is based on the presumption that κ is independent of temperature which is certainly not true for pure metals at low temperatures. Also, as κ increases at low temperatures, ΔT_J becomes small and difficult to measure precisely. For this reason pure copper was measured only down to liquid nitrogen temperature. Below this

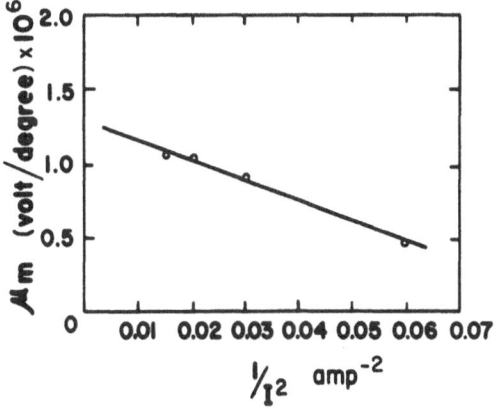

Fig. 3.6. Dependence of measured Thomson coefficient μ_m on the current I (from [28B1]).

temperature wires of copper and silver alloyed with gold were used. The aim was to add sufficient gold to make κ as independent of temperature as possible. Borelius *et al.* calculated this to be 0.37 at.% Au in both cases—and so copper-normal and silver-normal were born. In actual fact, the heat conductivity of the alloy wires still decreases significantly at the lowest temperatures.

The dominant source of error manifested itself in the dependence of the measured Thomson coefficient on current I, as shown in Fig. 3.6 for silver-normal. This was ascribed to the Peltier heat caused by slight inhomogeneities in the wire. The true Thomson coefficient was calculated from

$$\mu = \mu_m + A/I^2$$

where $\mu_{\dot{m}}$ is the measured Thomson coefficient and A is a constant. The basis for this equation is not discussed. The final corrected form of the Thomson coefficient is labeled 4 in Fig. 3.7.

As a check on their results, Borelius and his co-workers used

$$S_2 - S_1 = \int_{T_1}^{T_2} \left(\frac{\mu_1}{T} - \frac{\mu_2}{T} \right) dT \tag{3.10}$$

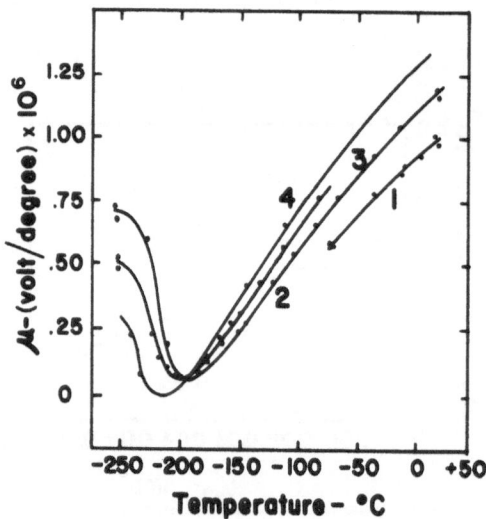

Fig. 3.7. Thomson coefficient for silver-normal. (1) $I = 7$ A. (2) $I = 10$ A. (3) $I = 14$ A. (4) Corrected to $1/I^2 = 0$ (from [28B1]).

Table 3.1.

	T_2, K	T_1, K	S_2-S_1, μV/K	
			from Eq. (3.10)	Measured
2 = silver-normal	17.3	70.6	−2.24	−2.18
	280.7	70.6	−2.01	−2.19
2 = copper	70.6	280.7	−2.53	−2.62

which is a combination of Eqs. (3.2) and (1.9), to calculate $S_2 - S_1$ from their measured values of μ_1 and μ_2. The subscript 1 refers to copper-normal and 2 to silver-normal or copper. $S_2 - S_1$ was also measured directly from thermocouples of the same materials. The results in Table 3.1 give an indication of the accuracy of the measurements. It is apparent that there are differences of about 10% in this table. In view of the discrepancies of about 6% in the thermoelectric comparisons of lead with silver-normal, it would seem that we might expect errors of about 15% or more in the absolute thermoelectric power of lead between 20 and 300 K.

Admittedly present-day theory cannot predict the thermopower even to this degree of accuracy. However, discrepancies of this order may considerably influence the apparent temperature dependence of the thermopower which is commonly relied upon to identify scattering processes. The situation is clearly unsatisfactory, and a redetermination of the absolute thermopower of lead using all the advantages of modern techniques is surely overdue.

3.5 Measurement of Unconventional Thermoelectric Coefficients

In Table 1.3 we listed various thermoelectric coefficients that can be defined through the use of the linear transport equations. In the past, those most commonly measured were σ, κ, and S. These can then be used to determine the quantities L_{ij} of Eqs. (1.15) and (1.16) or, equivalently, K_n defined by Eq. (2.23), thus uniquely giving these transport coefficients. With the advent of sophisticated measuring techniques which we shall discuss presently, one may well ask if it may be more convenient or instructive to measure other experimental quantities to determine K_n.

This question has been considered by Tracy and Smrčka [74T1], who conclude that of the twelve possible quantities in Table 1.3 four are

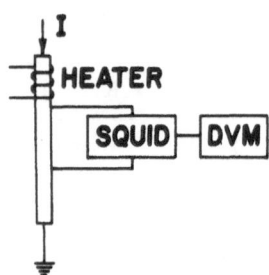

Fig. 3.8. Schematic arrangement for the measurement of the thermoelectric ratio, G, using a SQUID.

presently beyond reach, namely, Π_t,* Ω_t, S_a, and Σ_a. Σ_e could possibly be measured, and the remaining seven parameters can definitely be determined. Of these, three are properties traditionally investigated, leaving the remaining four, Π_e, κ_e, Ω_s, and σ_a.

Of these four, Π_e appears to be the most attractive, and it has, indeed, been studied recently by Garland [74G2], who measured $G = 1/\Pi_e$ for tungsten, and by Tracy [74T2], who studied silver. A schematic diagram of the apparatus is shown in Fig. 3.8, and a more detailed description is given in the paper by Garland, who also presents a careful thermodynamic analysis of the effect. Briefly, both a heat current and an electric current are passed through the sample, and the electric current is adjusted until the voltage across the sample vanishes. This voltage is monitored by a SQUID (see below).

As pointed out by Garland and by Tracy, G can be measured much more precisely at low temperatures than can S, and this parameter has the further advantage that, according to elementary theory, it should be independent of temperature as $T \to 0$ K, as contrasted to S, which should vanish as $T \to 0$ K.

We note in passing the following properties of G [74G2]:

$$G = eK_1 K_2^{-1} = S_e(e^2 K_0 K_2^{-1} T) \tag{3.11}$$

In the relaxation time approximation

$$G = e[\partial \ln \sigma(\varepsilon)/\partial \varepsilon]_\eta \tag{3.12}$$

and quite generally

$$G = e[\partial \ln \kappa_e(\varepsilon)/\partial \varepsilon]_\eta \tag{3.13}$$

---

* Π_t is just the Peltier coefficient which can be determined as described above. However, if one attempts to do so using the prescription of Table 1.3, i.e., monitoring ΔT, while maintaining it zero by balancing J and Q, various experimental difficulties arise.

where, however, κ_e is the thermal conductivity subject to the boundary condition $E = 0$, not $J = 0$, which is the more common.

If the relaxation time is the result of two independent scattering mechanisms, resulting in resistivity contributions ρ_1 and ρ_2,

$$G = \frac{G_1\rho_1 + G_2\rho_2}{\rho} \tag{3.14}$$

This result does not assume the validity of the Wiedemann–Franz law and, to this extent, is more general than the analogous Gorter–Nordheim rule for the TEP (see Chapter 2).

For a multiband model one obtains

$$G = (1/\kappa_e) \sum \kappa_{ej} G_j \tag{3.15}$$

where κ_{ej} is the thermal conductivity due to the jth band.

3.6 Measurement of Temperature and of Small Voltages

3.6a *Temperature Measurement*

Investigations of thermoelectric effects generally call for the measurement of temperatures, temperature differences, and rather small voltages. Since the techniques for the measurement of temperature and temperature differences have been discussed in a rather complete manner elsewhere [73I1], we shall make only a few brief comments. On the other hand, work in the past ten years has resulted in very significant advances in the measurement of small voltages. Since these results are as yet described only in the periodical literature, we shall devote this section largely to a description of the new devices and their application.

In the temperature region below about 15 K, germanium and carbon resistance thermometers are widely employed for temperature measurement. They are reliable, fairly compact, and have high sensitivity. Frequently, the sensing element is incorporated into an ac bridge, and the imbalance of the bridge is observed with a phase-sensitive detector. Above about 15 K, the platinum resistance thermometer is the accepted standard. Although reliable and simple to use, this unit has the disadvantage of fairly large size.

For many purposes, thermocouples are to be preferred. Their principal drawbacks are in the low-temperature region (<1 K for Au–Fe thermocouples), where the sensitivity is rather poor and where, moreover, nominally identical thermocouple wires from different suppliers may have slightly different characteristics. For precise work with thermocouples, especially the gold–iron vs. chromel couples, it is probably best to calibrate the couple against a platinum resistance thermometer in the higher temperature region, and against a calibrated germanium resistance thermometer at low temperatures. The other disadvantage of thermocouples has been alluded to earlier; spurious thermals can cause trouble, especially when the hot and cold junctions are at temperatures near 4 K and the leads to the measuring instrument therefore traverse a very large temperature difference. In this situation it is important that some care be exercised to minimize these spurious thermal voltages and that occasional checks be performed to see that they are, indeed, negligible. On the other hand, thermocouples have the great advantage that they can be used to measure the temperature of a very small region, that of the "hot" junction (assuming the cold junction is immersed in the cryogenic fluid); have a very small heat capacity and, consequently, respond rapidly to changes in temperature; require no external source of power; and can be constructed of sufficiently fine wires so that the thermal conductance between the two junctions is negligibly small. Hence, except for calibration runs or the measurement of very small temperature differences at low temperatures, thermocouples are generally satisfactory and convenient sensors.

3.6b *Voltage Measurement*

As the temperature is lowered toward absolute zero, the TEP must also approach zero. In the vicinity of 4.2 K and below, the TEP of metals and alloys is generally very small, of order 10^{-8} V/K or less. The use of either the integral or incremental method then requires accurate determinations of very small voltages and, consequently, very sensitive devices and also cunning schemes to eliminate or at least greatly reduce spurious thermal emfs.

Prior to about 1965 good potentiometers were the essential voltage-measuring devices in the laboratory. Today, potentiometers have been virtually displaced in research laboratories first by analog micro- and nanovoltmeters, and more recently by the much more rapid and convenient digital voltmeters, whose range now extends into the nanovolt region. The latter instruments have very high common mode rejection ratios which eliminate most of the grounding problems that have plagued the users of analog meters. The digital meters also have the further advantage that their outputs may be fed directly into a data recording system or to a minicomputer for on-line data analysis.

Frequently, however, the use of sensitive digital voltmeters is insufficient to the task. Over the years, various ingenious devices, many employing superconducting elements, have been developed, and one of these, the SQUID (superconducting quantum interference detector), is now commercially available and widely used.

Before discussing the superconducting devices, we shall mention briefly two techniques using conventional circuitry that have been employed with some success. The first is a magnetic amplifier developed by Foiles [67F2] and illustrated in Fig. 3.9. An audio-oscillator feeds two coils, wound in opposition, on either side of the central I of the transformer. The output coil wound on the central I of the transformer is fed to a phase-sensitive detector. The unknown emf is placed in series with a known small standard resistor and supplies a current through the control winding, which is also wound about the central transformer section.

In the absence of a current through the control winding, the output to the PSD is zero. When a current flows through the control winding, there is

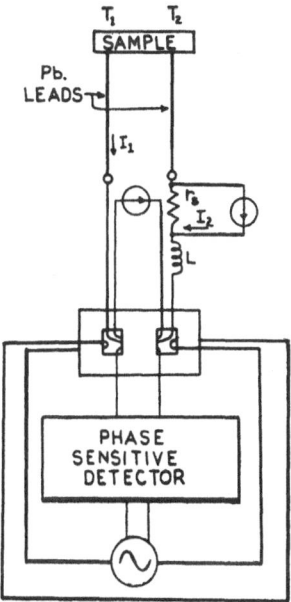

Fig. 3.9. Schematic diagram of a magnetic amplifier (from [67F2]).

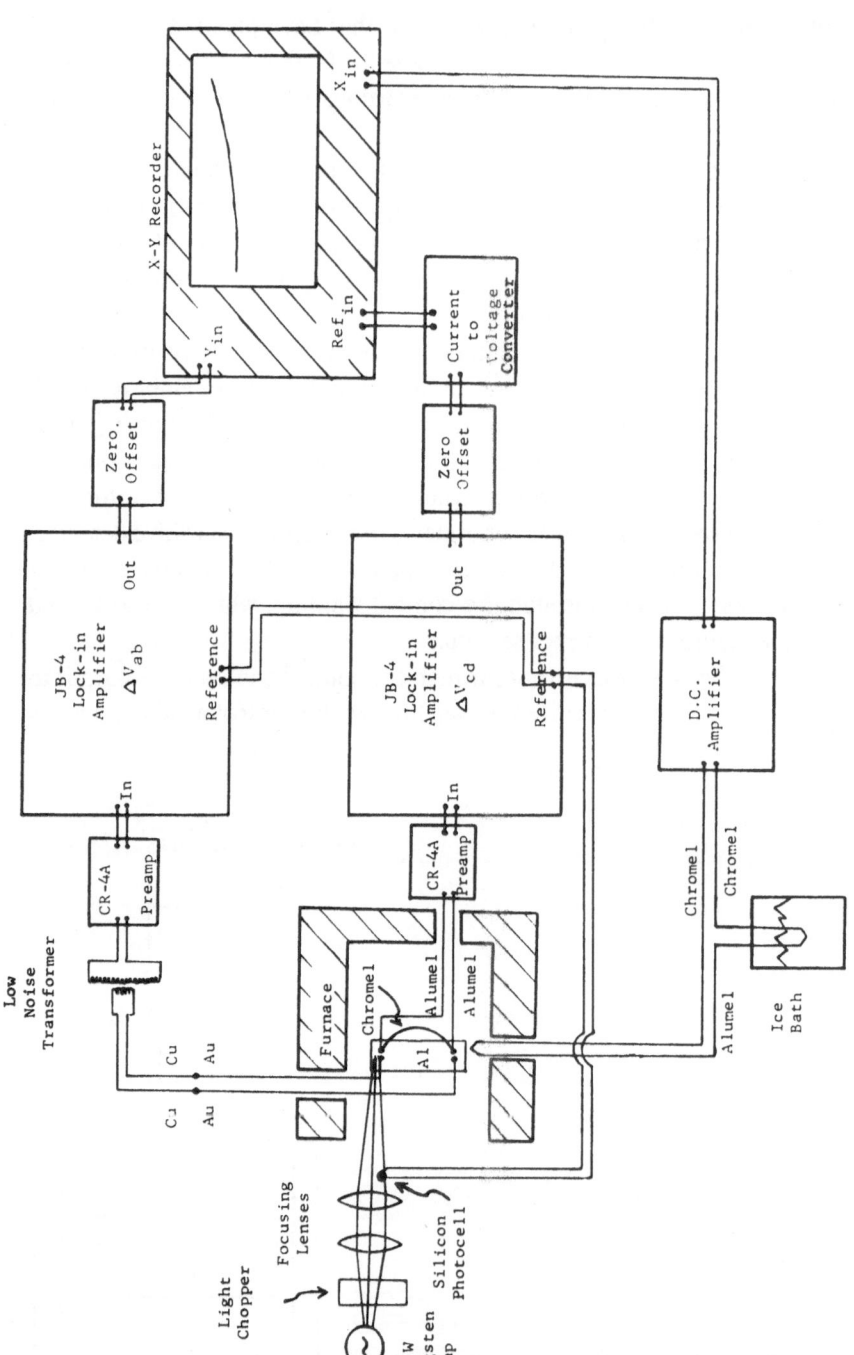

Fig. 3.10a. Schematic diagram of the system for measuring thermopowers by temperature modulation (from [70F1]).

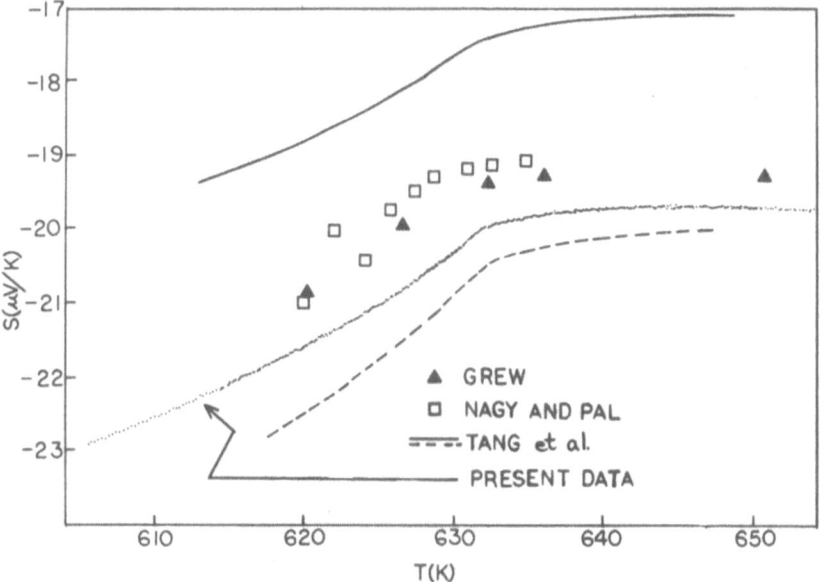

Fig. 3.10b. The thermopower of nickel near the Curie temperature; data obtained with the system shown in Fig. 3.10a. The solid and dashed curves are from measurements by Tang *et al.* [71T1, 74T3]; results of Grew [32G1] and of Nagy and Pal [70N4] are also shown (from [74P1]).

also an output to the PSD. During operation, the unknown emf is bucked by passing a current through the small resistor r_s until the PSD reads zero. With a source impedance of 0.015 Ω, the device has a sensitivity of 5 nV; Foiles mentions a number of ways in which this can be further improved. One of the principal advantages of the device is that spurious thermal voltages are effectively eliminated since the transformer is submerged in the helium bath.

Another interesting method for overcoming the effects of dc thermal emfs in the leads has been developed by Freeman and Bass [70F1]. In their apparatus the temperature of one junction is modulated by chopping a light beam focused on the junction, thereby producing a modulated thermal emf which could be detected using lock-in amplifiers, as shown in Fig. 3.10a.

If a small temperature difference ΔT is maintained between the junctions of a thermocouple composed of metals A and B (Au and Al in Fig. 3.10), the resulting voltage is

$$\Delta V_{AB} = (S_B - S_A)\Delta T$$

where S_A and S_B are the TEPs of metals A and B. If the same temperature difference is also maintained across the junctions of a second thermocouple whose thermopower S_{CD} is known, then

$$\Delta V_{CD} = S_{CD} \Delta T$$

and

$$\frac{\Delta V_{AB}}{\Delta V_{CD}} = \frac{S_B - S_A}{S_{CD}}$$

Consequently, if $\Delta V_{AB}/\Delta V_{CD}$ and the average temperature of the thermocouples are measured, and S_{CD} and S_B are known, the TEP of the unknown S_A, can be determined without an explicit measurement of ΔT.

The technique was used by Piotrowski et al. [74P1] to measure the thermopower of nickel in the vicinity of its Curie temperature. The thermocouple voltages were fed into two lock-in amplifiers with the same reference signal derived from the light chopper, and the ratio of the amplifier outputs was measured by a digital multimeter in the ratio mode. Metals C and D were gold and platinum, and S_{CD} was obtained from independent measurements using standard techniques. Metals A and B were nickel and platinum. One of the major advantages of this method, in addition to eliminating spurious dc thermal voltages, is that the temperature differences across the thermocouple junctions can be kept quite small, about 0.05 K, a particularly important consideration for work in the vicinity of a phase transition. In Fig. 3.10b we show the results for nickel, consisting of some 500 data points over a temperature range of about 50 K near $T_c = 631$ K.

3.7   Superconducting Devices

The superconducting devices fall into two categories. In the first, the superconductor is used to modulate the dc signal at the source. This permits the application of ac amplification techniques to the modulated signal and, perhaps more important, effectively suppresses spurious thermal emfs originating in the leads from the cryostat. In the second group, the devices are based on the characteristics of Josephson junctions or weak links and sense the magnetic field due to the current in a closed circuit containing the unknown voltage source. Generally, these devices are used in a null-detector configuration, and in most applications their sensitivity is limited by Johnson noise.

Historically, the first superconducting device falls into neither of the above categories. This device was a superconducting galvanometer which was developed by Pippard and Pullan [52P1] following an earlier design by Casimir and Rademakers [47C1]. The unit was capable of detecting 10^{-12} V in circuits of about 10^{-7} Ω resistance. The rather long periods (~12 sec), the fragile nature of the quartz suspension, and the somewhat clumsy bulkiness of the galvanometer have mitigated against its wide use.

3.7a Superconducting Modulators

The device constructed by the low-temperature group at the National Research Council, Ottawa, is shown schematically in Fig. 3.11. Resistors R_1, R_1', R_2, and R_2' are short sections of superconducting niobium wire. Each of the wires was surrounded by a small coil of wire through which a current of sufficient magnitude could be passed such that the resulting magnetic field exceeded the critical field of niobium. When current is passed through the coils surrounding R_2 and R_2', the voltage between points B and D is the voltage of the source. The voltage between points B and D is reversed by passing a current through the coils around R_1 and R_1', and setting that in the coils of R_2 and R_2' to zero. Since the source voltage is alternately shunted by the parallel combination of R_1 and R_1' or R_2 and R_2', the use of this reversing switch is limited to voltage sources with resistance much less than these resistances.

At Ottawa this device was used as a reversing switch in conjunction with a very sensitive photoelectric amplifying galvanometer in the classic determination of the absolute thermoelectric power of lead below 18 K [58C1]. Similar devices have been used by Guénault [67G1, 74G5] for measurements of the TEP of noble metals and their alloys. However, the device could also be employed as a modulator by simply switching the coil currents at a fixed low frequency.

In the modulating mode this device is not as useful as those developed by Templeton [55T1] and de Vroomen and van Baarle [57D1]. In these, modulation is achieved by placing a section of superconducting wire in series

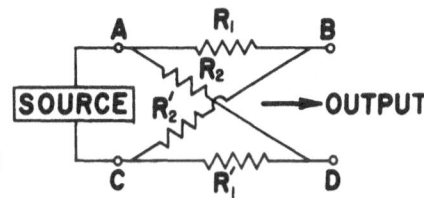

Fig. 3.11. Schematic of superconducting reversing switch.

with the unknown voltage source and rendering this wire normal at regular intervals by an alternating magnetic field produced by a coil surrounding the superconductor. The alternating current thus produced passes through a transformer to the electronic detection system. One of the problems encountered with such modulators is that of large stray pickup from the modulating coil. In the case of the latter modulators, this problem is relatively small because a modulating current of frequency f in the coil renders the superconducting wire normal twice each period. Hence, the desired signal voltage has a frequency $2f$, and the undesired pickup can be suppressed by means of a tuned amplifier. On the other hand, periodic switching of the reversing switch results in a pickup of the same frequency as the modulated signal.

One of the drawbacks of these modulators is that they are difficult to employ in a magnetic field because of the elaborate shielding precautions required. Even then, their use is restricted to relatively low external fields. This restriction was overcome by Edwards [71E1], who used a type II superconductor with high critical field (niobium–zirconium) and accomplished the switching thermally by application of an alternating current to a small heater surrounding the superconductor. A schematic of the circuit is shown in Fig. 3.12. As for the magnetically activated modulator, the modulation frequency is $2f$, where f is the frequency of the current supplied to the heating element. Pickup is further reduced by noninductive winding of the heater. Generally, the apparatus is used as a null detection system; the current passing through the reference resistor is adjusted until zero output is obtained, at which point the voltage across the reference resistor just balances that across the sample.

Edwards employed the device in a study of the magnetoresistance of single-crystal indium, but it has also been used by Averback and Bass [71A1] in measurements of thermoelectric voltages in a magnetic field. Normally, the device is operated at a heater current frequency of 200 Hz, though it can operate at heater frequencies up to the surprisingly high value of 1000 Hz. The sensitivity is typically 10^{-11} V at zero field and 10^{-9} V at 100 kG. In the original model the resistance of the superconducting switch was $0.02\ \Omega$, so that for efficient use the specimen resistance should be substantially smaller than this. Allnut and Walton [72A1] have constructed a similar device using 60 turns of lead-coated 50 swg constantan wire. This has the advantage that the transition temperature (7.2 K) is considerably lower than that of Nb–Zr (10.8 K), and the unit therefore requires much less heat input. Moreover, with their design a normal-state resistance of the

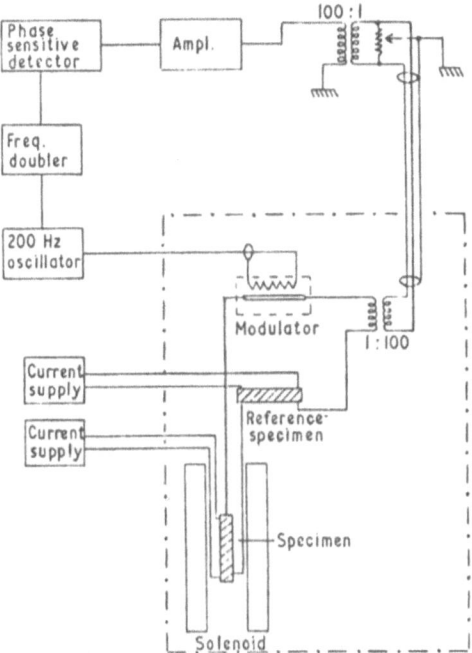

Fig. 3.12. Schematic circuit diagram of the super-conducting chopper amplifier. All components inside the dash-dotted rectangle are in the helium bath.

modulator of about 100 Ω was achieved. The principal disadvantage is that this device cannot be used conveniently in a magnetic field because the lead coating is then rendered permanently normal.

3.7b Weak Link and Josephson Junction Devices—SQUIDS and Slugs

3.7b1 SQUIDS [72G3]

To illustrate the operating principles of a SQUID (superconducting quantum interference detector) [70Z1], we consider first the characteristics of a simple superconducting loop containing a weak link W, as shown in Fig. 3.13. A "weak link" is essentially a highly constricted region, for example, a point contact. For our purposes the important property of such a weak link is its relatively low critical current.

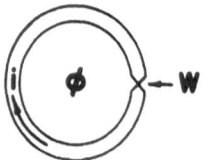

Fig. 3.13. Superconducting loop with a "weak link" at W.

Let us assume that initially the region containing the loop is flux-free and that the loop carries no current. If we now apply an external flux, a current is induced in the coil which is such as to cancel exactly the external flux in the loop, maintaining zero flux within the loop. As the external flux ϕ_x is increased, the loop current i also increases until the critical current i_c of the weak link is reached. As the link W becomes normal, one flux quantum $\phi_0 = hc/2e = 2 \times 10^{-7}$ G-cm$^2 = 2 \times 10^{-15}$ Wb enters the loop. Simultaneously, the current i diminishes since the difference between ϕ_x and the flux within the loop has been reduced by ϕ_0. If the external flux is now increased further, W becomes normal once again, admitting another flux quantum, and so on. The resulting characteristic of the loop is then like that of Fig. 3.14. The initial value of $k = \phi_x/\phi_0$ for the first flux jump, and current discontinuity may be quite large. Thereafter, however, transitions occur at increments of $\Delta k = 1$. Note also that the characteristic is not symmetrical: Suppose point H has been reached in the ascent, and ϕ_x is now reduced; the reverse transition HG does *not* take place since i is now at a value

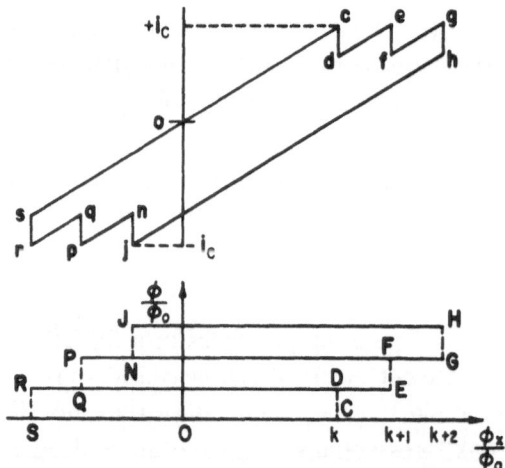

Fig. 3.14. Flux and induced current in superconducting loop with weak link, as function of excited flux.

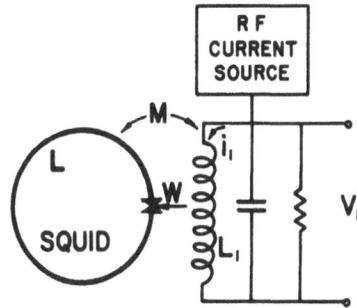

Fig. 3.15. Elementary rf SQUID circuit

appropriate to the flux corresponding to point H. Only after ϕ_x/ϕ_0 has diminished by $2k$ will $i \to -i_c$ and the transition JN occur.

The flux path $OCDEFGHJNPQRS$ and the corresponding current path $ocdefghjnpqrs$ are the static characteristic of the loop over the interval shown. The SQUID is an rf device (it may be called a stable, rf-biased, superconducting, point contact device), and we shall assume that the same characteristic applies when ϕ_x is varying sinusoidally at radio frequencies.

Figure 3.15 shows the elements of the rf SQUID circuit, consisting of a tuned tank circuit inductively coupled (through a mutual inductance M) to the loop L containing the weak link W. If we denote the tank circuit current by i_1, then $\phi_x = Mi_1$, and Fig. 3.14 could be redrawn with i_1 replacing ϕ_x, as shown in Fig. 3.16.

The coil current i_1 consists of an average dc component i_0 and a superimposed rf component i_r. We shall first consider two limiting cases, when i_0 has the values i_{0A} and i_{0B}, as shown in Fig. 3.16. Points A and B are rather special points situated midway between NG and PG, respectively.

If $i_0 = i_{0A}$, then for small values of i_r the flux ϕ through the SQUID loop maintains itself, ideally, at some constant value. If the amplitude of i_r is increased (curve b in Fig. 3.16), point G is reached and a transition to H occurs. On the reverse part of the cycle the path $HJNA$ is followed. Since the loop $AGHJNA$ has the dimensions of energy, we have here a hysteresis effect which entails absorption of energy from the primary tank circuit. We shall consider the implications of this hysteresis shortly, but first we shall follow the flux pattern as the amplitude of i_r is increased further. A small increase in i_r does not result in an increase in hysteresis, but at some point we shall follow the path $AGHT$ on the positive cycle, resulting in a transition to point U. The remainder of the cycle is then $VYJNPQEF$, and the path now encloses three elementary hysteresis loops. Similarly, as i_r is increased

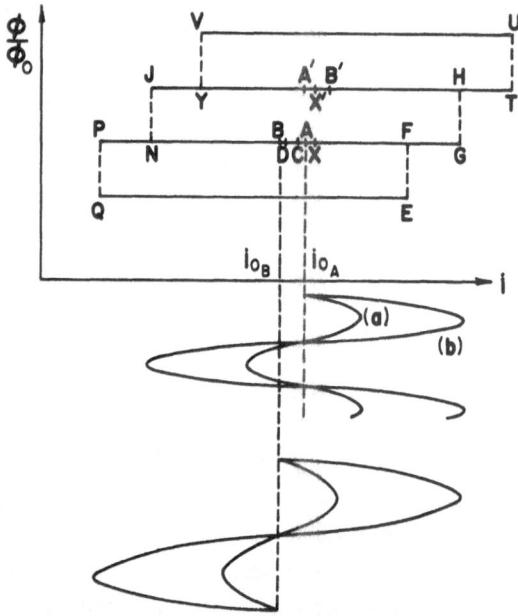

Fig. 3.16. Flux pattern in SQUID loop as a function of
rf and bias currents.

further, we shall always enclose an odd number of hysteresis loops, provided
$i_0 = i_{0A}$.

If the biasing current is set to i_{0B}, no flux jump occurs until point G is
reached. Then the path is *BGHJNPQEF*, enclosing two elementary loops.
The use of higher rf amplitudes will result in enclosing an increased even
number of loops. We can now understand qualitatively the experimentally
observed oscillograms of the rf voltage V_1, the voltage across the primary
coil, as i_r is changed.

We first assume that $i_0 = i_{0A}$. For small values of i_r, the system is lossless
and V_1 is proportional to i_r. However, when point G is reached, energy is
absorbed from the tank circuit, and V_1 cannot increase above the value at
which the transition occurs until the rf excitation is sufficient to compensate
within each cycle the energy lost in going around the hysteresis loop. Hence,
the pattern follows the path *ab* in Fig. 3.17. Thereafter, with further increase
in rf excitation, V_1 will increase again along *bc*. Point *c* corresponds to point
T of Fig. 3.16. At this point, a triple hysteresis loop is excited, and V_1 again
remains at a constant level until the energy loss per triple loop is replaced in

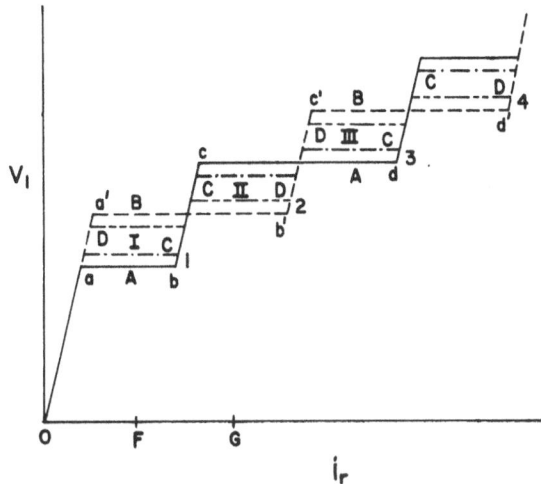

Fig. 3.17. Tuned circuit voltage, V_1, of Fig. 3.15, as a function of rf source current. A, B, C, D refer to different values of bias current, as indicated in Fig. 3.16. I, II, III, ... indicate the number of loops enclosed. This is the so-called step pattern of the SQUID. ——— path A; - - - path B; - - - path C; - - - - path D.

one rf cycle. Thus, the plateau cd will be somewhat longer than ab. Thereafter, the pattern continues with plateaus of equal length.

If the bias current is i_{0B}, the first plateau will appear at a somewhat higher excitation a', and the lengths of the plateaus $a'b'$, $c'd'$, etc., will be equal since in each instance two additional elementary hysteresis loops are excited.

As we pointed out above, bias currents i_{0A} and i_{0B} correspond to rather special points on the current–voltage characteristic. Let us now consider what might happen if instead of varying the rf excitation we maintain it constant but change i_0, the biasing current. To do this, we first consider the form of the V_1–i_r characteristic, if i_0 assumes some value other than i_{0A} or i_{0B}. For example, if $i_0 = i_{0C}$, then the characteristic will be that labeled C in Fig. 3.17. Two points should be noted: first, the rf amplitude for which one hysteresis loop is traced out is now greater than it was when $i = i_{0A}$, but less than when $i_0 = i_{0B}$. Second, before point c (Fig. 3.17) can be reached, the two loop path $AGHJNPQEF$ will be traced through, and hence the characteristic will fall below cd, the plateau corresponding to the excitation of three hysteresis loops. Similarly, if $i_0 = i_{0D}$ the characteristic will be that designated by D in Fig. 3.17. In fact, further consideration of these characteristics

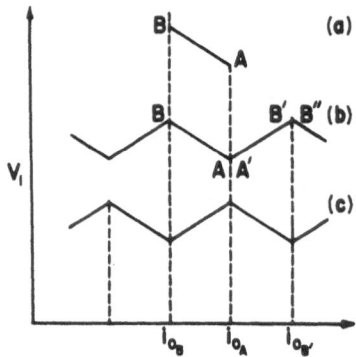

Fig. 3.18. Tuned circuit voltage, V_1, of Fig. 3.15 as a function of bias current with constant rf excitation. This is the so-called triangle pattern of the SQUID. (a) Characteristic for i_0 changing between i_{0A} and i_{0B} (Fig. 3.16) and i_r fixed at F (Fig. 3.17). (b) Complete characteristic for i_r at same value, but with i_0 ranging beyond i_{0A} and i_{0B}. i_{0x} is obtained from $i_{0x} = i_{0x''}$, where $i_{0A} - i_{0x''} = i_{0x} - i_{0A}$. (c) Characteristic for i_r corresponding to G (Fig. 3.17).

leads to the conclusion that one hysteresis loop of Fig. 3.16 is traced out for all V_1–i_r points within the area labeled I in Fig. 3.17, two hysteresis loops are traced out for all points within the area II, and so on.

Suppose now that i_r is maintained constant, corresponding to the amplitude OF of Fig. 3.17, but that i_0 is varied periodically between the limits i_{0A} and i_{0B} at audio frequency. The appropriate V_1–i_0 characteristics are shown in Fig. 3.18a. To see what would happen if i_0 falls outside these limits, having the value i_{0x}, for example, we return to Fig. 3.16. If the rf amplitude is XG, an upward flux jump will take place, but the amplitude of the R.F. current is then insufficient to cause the downward transition JN. The flux will then maintain itself at the value corresponding to the line JT, centered at point X'. This point is between points B' and A', which are defined similarly to A and B. If we now increase the rf amplitude further, we will get a downward transition and trace out the hysteresis loop $GHJN$. This response is the same as that obtained for a bias current $i_{0x''}$, where $AX = A'X' = AX''$ (Fig. 3.16). This identifies V_1 corresponding to i_{0x} in Fig. 3.18, and the remainder of the $A'B'$ branch is obtained in a similar fashion. Hence the V_1–i_0 characteristic will exhibit the repetitive triangular pattern shown in Fig. 3.18b. If the rf amplitude is OG in Fig. 3.17, similar reasoning shows that i_{0B} corresponds to a minimum in V_1 and i_{0A} to a maximum, and the corresponding characteristic is that shown in Fig. 3.18c.

These patterns can be displayed on an oscilloscope using the circuitry of Fig. 3.19. If the switch is in the "Steps" position, the rf oscillator amplitude is modulated by the audio oscillator. Using the audio oscillator output as the horizontal signal on the scope gives a deflection proportional to the rf amplitude, and the pattern on the scope will then correspond to Fig. 3.17. If the switch is in the "Triangles" position, the audio oscillator provides a

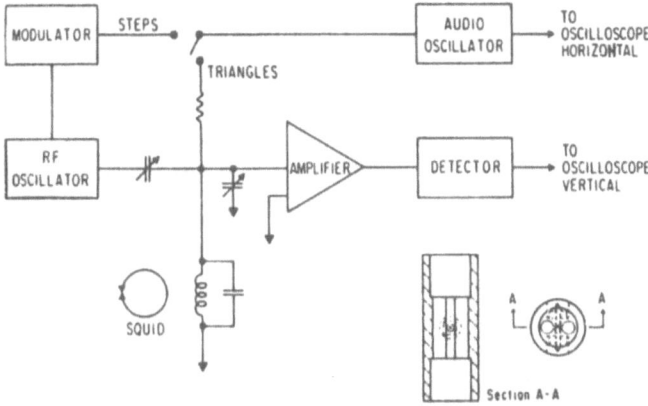

Fig. 3.19. Block diagram of circuit for observing step and triangle patterns of SQUID (from [72G3]).

time-varying bias current to the rf coil, and a pattern such as Fig. 3.18b will appear on the scope.

Having examined in some detail the behavior of a SQUID, we now see how the device can be used to measure very small currents. In practice, currents are measured by passing them through a small coil placed inside the SQUID loop, controlling the "external" flux. Suppose we set up the circuit of Fig. 3.19 with the "sample coil" coupled to the SQUID loop, and the switch in the "Triangles" position. We adjust the rf and audio amplitudes until the pattern BAB' of Fig. 3.18b is seen on the oscilloscope. Recalling that the variation of V_1 with i_0 in Fig. 3.18 is the result of the flux produced by i_0 which links the SQUID ring, we see immediately that if we send a current through the sample coil, its effect will be to displace the oscilloscope pattern horizontally, just as though we had changed the average bias current. In particular, if the flux through the SQUID changes by one flux quantum, the pattern will be displaced by one period, i.e., B'' will appear at the position of B in Fig. 3.18. Flux changes can therefore be measured by counting the number of triangles which pass a given point on the oscilloscope screen.

Generally, however, it is more convenient to connect the output of the detector of Fig. 3.19 to a lock-in amplifier which is synchronized with the audio oscillator. A current proportional to the amplifier output is then fed back into the rf coil so as to compensate for the flux change induced by the sample coil in the SQUID. A voltage proportional to this feedback current can be monitored, and it is, of course, a direct measure of the current flowing in the sample coil.

An arrangement such as that just described has been used by Tracy [74T2, 74T4] to measure the thermopower of samples of $AuIn_2$ and of Ag. The thermo-emf forced a current through the sample coil in the SQUID and a series standard resistor R_{ST} of small value. The circuitry was arranged so that an externally controlled current could also be passed through R_{ST}. In the experiment, the internal feedback voltage is recorded when the temperature difference across the sample thermocouple is zero. One junction is then heated, causing a thermo-emf and resulting in a change of the internal feedback voltage. The external bucking current is manually adjusted until the feedback voltage returns to its original value. The thermoelectric voltage is then $i_F R_{ST}$, where i_F is the external bucking current. Both the internal feedback voltage and i_F are measured to high resolution using digital voltmeters.

With this arrangement 1 μA of current through the sample coil resulted in a flux change of about one flux quantum in the SQUID. The noise in the SQUID electronics corresponds to fluctuations of about $10^{-3}\phi_0$ in the SQUID. The current sensitivity is, therefore, of order 10^{-9} A, which in a circuit of 5×10^{-6} ohm gives a voltage sensitivity of 5×10^{-15} V. However, the sensitivity is limited by Johnson noise to about 5×10^{-14} V. Results obtained by Tracy for G using the method of Fig. 3.8 are shown in Fig. 3.20.

We note in passing that the SQUID used by Tracy is a symmetric double-hole device [66Z1]. The principal advantage of such a system is that

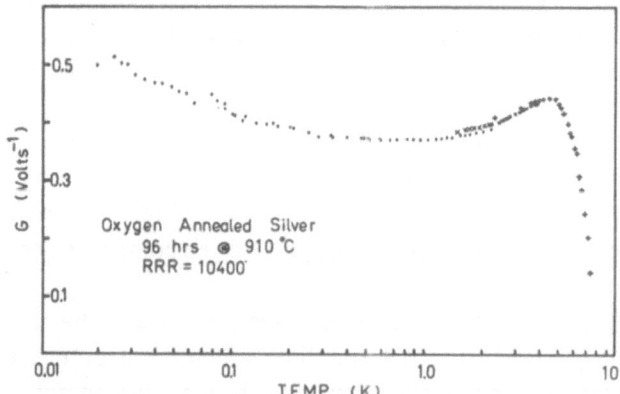

Fig. 3.20. The thermoelectric ratio G of pure silver at low temperatures. (Units are ampere/watt = volt^{-1}.)

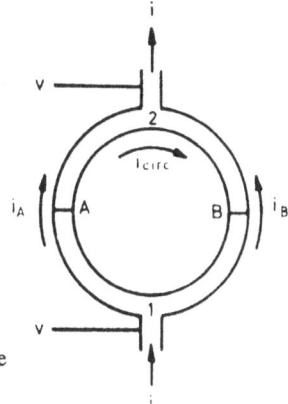

Fig. 3.21. Schematic diagram of a double-junction device (from [66C1]).

it is sensitive to external magnetic field *gradients* rather than the value of the fields. Thus, it is relatively insensitive to stray magnetic fields.

3.7b2 Slugs

Slugs, a name applied by Clarke [66C1], who developed them, are double-junction Josephson devices. The elements of the device may be understood by reference to Fig. 3.21. C and D are two superconductors separated by two Josephson junctions at A and B. The Josephson junctions are very thin insulating barriers separating the superconductors. When the current passing through the link is less than a well-defined critical current, no voltage appears across the barrier. When the current exceeds the critical value, the link ceases to be superconducting and a voltage appears across the barrier. For the parallel arrangement of Fig. 3.21 the critical current of the pair is a periodic function of the magnetic flux ϕ threading the ring, the modulation period being one flux quantum [64Z1], as shown in Fig. 3.22. If we now assume that the external flux, produced by a coil through which we pass a current i_H, is kept constant while we modulate the current i sinusoidally in such a way that the critical current is exceeded over some portion of

Fig. 3.22. Critical current of double junction as function of externally applied flux. ϕ is presumed to derive from a current i_H in a coil coupled to the double-junction loop.

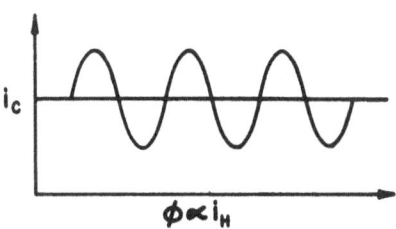

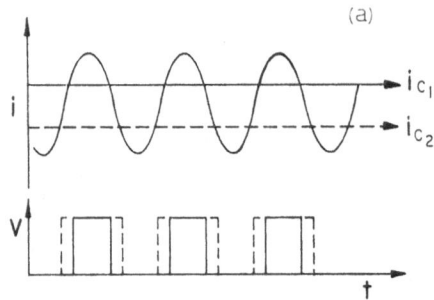

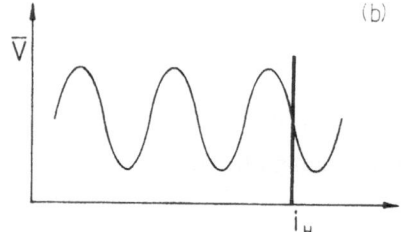

Fig. 3.23. (a) Voltage V from saturating amplifier connected to double junction as the current is modulated sinusoidally. (b) Time average output voltage of saturating amplifier as a function of i_H.

each cycle, then the voltage across the ring will be finite for this same portion of each cycle. If the voltage across the ring is fed into a saturating amplifier, the output will be pulses whose width will depend on the critical current and, hence, on i_H (Fig. 3.23a). If we vary i_H slowly, so that i_c varies periodically between the limits i_{c1} and i_{c2}, then, clearly, the pulse width and hence the average value of the amplifier output will also vary periodically, and we shall obtain the curve of $\bar{V}$ vs. i_H, such as shown in Fig. 3.23b.

To use the device as a null detector, we place it into a circuit such as shown in Fig. 3.24. Here E_x is the unknown voltage and R_s is a small standard resistance through which a current i_s can be passed so as to produce a voltage drop opposing E_x. Initially we assume that E_x and i_s are both zero and adjust i_H to a value shown by the dashed line in Fig. 3.23b; for this value of i_H, $d\bar{V}/di_H$ is a maximum. E_x and i_s are now increased so as to maintain $\bar{V}$ constant; E_x is then given by $i_s R_s$, where i_s is measured potentiometrically and R_s is known from prior calibration.

The sensitivity of such a circuit can be estimated as follows. We presume that the coil and ring are tightly coupled so that the mutual inductance is roughly equal to the self-inductance of the coil L_0. The current change required in L_0 to produce a flux change of one flux quantum in the ring is then approximately ϕ_0/L_0. Hence, the voltage E_0 required for one oscillation in the $\bar{V} - i_H$ characteristic of Fig. 3.23b is given by $E_0 = \phi_0 R/L_0$,

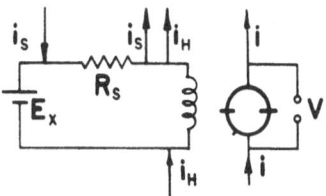

Fig. 3.24. Schematic of circuit using a "slug" as a null detector.

where R is the total resistance of the primary circuit. In terms of the time constant τ of the RL circuit, $E_0 = \phi_0/\tau$, i.e., for a time constant of 1 sec, $E_0 \sim 10^{-15}$ V.

If the number of oscillations of $\bar{V}$ is measured as the unknown voltage is applied, we have a digital voltmeter with a resolution of approximately 10^{-15} V. Alternatively, using the device as a null detector as indicated above, and assuming that changes of $\bar{V}$ equal to 1% of its periodic amplitude can be observed, in principle we have a null detector with ultimate sensitivity of about 10^{-17} V. In practice Clarke's slug described below had a current sensitivity of 1 μA corresponding to a sensitivity of 10^{-14} V for a circuit resistance of 10^{-8} Ω.

The actual device used by Clarke, consisting of a bead of solder on a short length of 0.004-in.-diameter niobium wire, bears little superficial resemblance to the schematic slug of Fig. 3.21. Clarke suggests the following

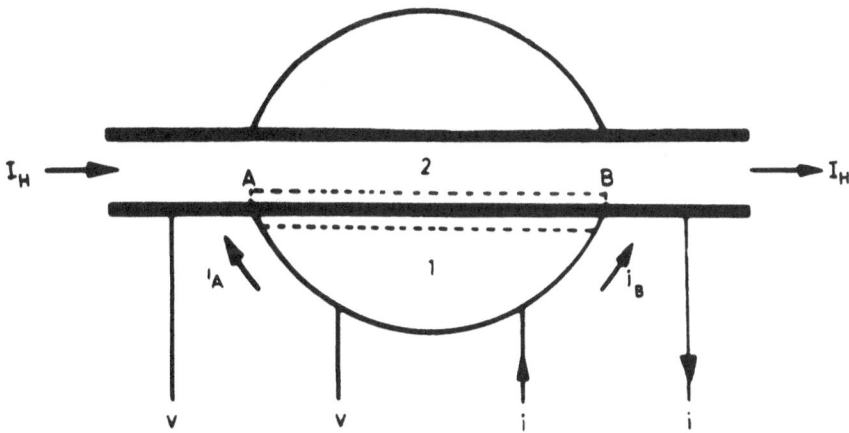

Fig. 3.25. Actual "slug" consisting of niobium wire and solder bead. The heavy lines represent the insulating layer of niobium oxide and the dotted lines enclose the region into which the flux penetrates. Assuming the junctions are formed at A and B, this "slug" is identical to Fig. 3.21 (from [66C1]).

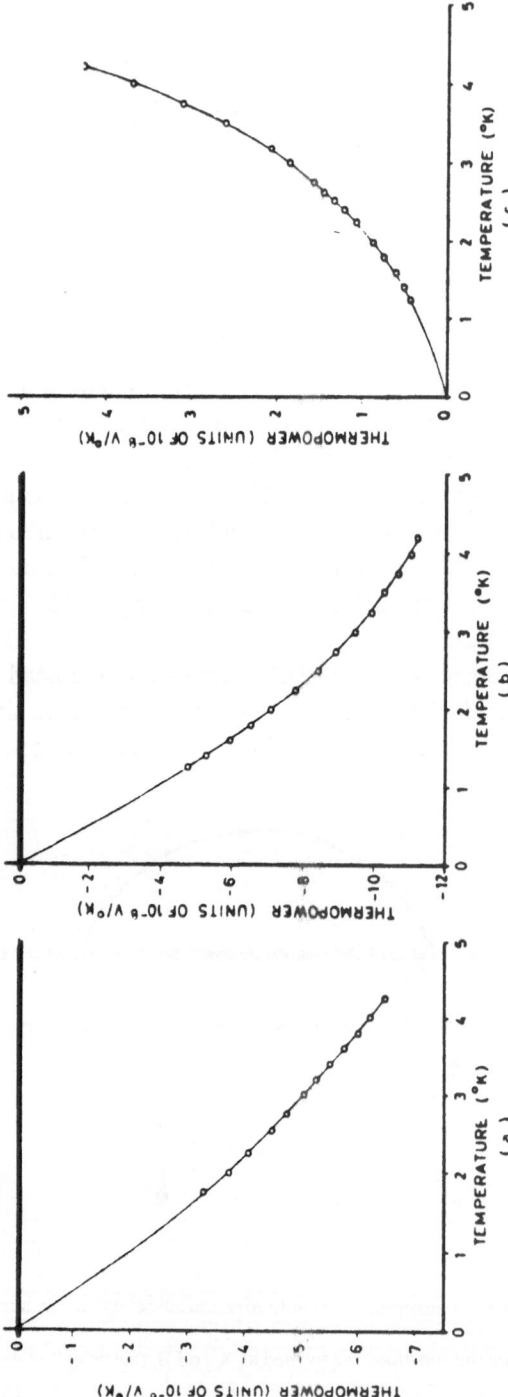

Fig. 3.26. Low-temperature thermopower: (a) single crystal of copper; (b) same crystal after oxygen anneal; (c) polycrystalline 6N pure silver (from [69R1]).

Table 3.2. Methods Used in Thermoelectric Measurements

Method	Current sensitivity	Voltage sensitivity			Use in magnetic field	Temperature range
		$R = 10^{-6}\,\Omega$	$R = 10^{-4}\,\Omega$	$R = 1\,\Omega$		
DVM or potentiometer + mechanical reversing switch		$\sim 10^{-8}$ V	$\sim 10^{-8}$ V	$\sim 10^{-8}$ V	Yes	$\sim 4.2 \rightarrow$
DVM or potentiometer + superconducting reversing switch		$\sim 10^{-8}$ V	$\sim 10^{-8}$ V	$< 10^{-8}$ V	No	< 20 K
Magnetic amplifier [67F2]	3×10^{-7} A	3×10^{-13} V	3×10^{-11} V	3×10^{-7} V	Up to 10 kg with shielding	< 25 K
Superconducting modulator [71E1]	2×10^{-7} A	2×10^{-13} V	2×10^{-11} V	2×10^{-7} V	Yes	< 20 K
Peltier heat equivalent [69T1]		1.5×10^{-11} V	1.5×10^{-11} V	1.5×10^{-11} V	Yes	< 20 K
Slug [66C1]	10^{-6} A	10^{-12} V	(2 K) 10^{-10} V	(2 K) 10^{-6} V	Not yet	< 10 K
SQUID [74T2]	10^{-9} A	10^{-15} V	10^{-13} V	10^{-9} V	Not yet	< 10 K
Johnson noise at 4.2 $V/Hz^{1/2}$		1.5×10^{-14} V	1.5×10^{-13} V	1.5×10^{-11} V		

mechanism (see Fig. 3.25). Superconductors of the ring are the solder bead and niobium wire. The device behaves as though the weak links exist only at each end of the solder, at A and B. The area between the dotted lines which are separated by the penetration depths of the niobium and solder is equivalent to the area of the ring, and the current flowing in the niobium wire produces the flux ϕ which passes through this area. Clarke also describes a modified version in which a small incision is made in the coating of a 0.002-in.-diameter niobium wire; the wire is then immersed in solder as before.

Such slugs have been used successfully by several workers. Rumbo [69R1] has measured the thermopower of single crystals of copper and of polycrystalline silver. His results on copper are shown in Fig. 3.26. Besides illustrating the effect of removal of iron from solid solution by oxidation, these curves give some indication of the precision attainable with these devices at low temperatures. Garland and Van Harlingen used a slug in their measurement of the thermoelectric ratio G, which we mentioned earlier [74G2]. Other improvements and modifications, including feedback systems similar to those for the SQUID, are described by Zych [68Z1] and Smith and Anderson [71S1]. With the advent of SQUIDS and slugs the temperature region below 1 K has become accessible for thermoelectric measurements, and we can look forward to increasing research activity here.

In Table 3.2 we summarize some of the properties of the methods we have studied. It becomes clear that the current devices—magnetic amplifiers, superconducting modulators, slugs, and SQUIDS—can have extremely high sensitivities, possibly restricted by Johnson noise, for very low circuit resistances. They are therefore ideal for studies on pure metals at low temperatures. For alloy samples where the resistances may be high, their voltage sensitivities become comparable or even less than those of the voltage devices—digital voltmeters and potentiometers.

4 | Phonon Drag

4.1 Introduction and General Relations

The topic of this chapter has already been introduced in Chapter 2. There we used plausibility arguments to derive an elementary expression for the phonon-drag thermopower S_g, and presented qualitative arguments to explain some of the dominant features of this phenomenon. We shall now carry the discussion much further, describing in greater detail both theory and experimental observation. The reader will be aware of some degree of overlap between certain portions of this chapter and of Chapter 2. We trust that the redundancy will obviate the need for excessive page turning.

In Chapter 2 we obtained an expression for S_g, Eq. (2.35). We shall present shortly the results of the formal theory; however, it will assist in understanding the origins of the various terms of that theory to pursue this simplified model briefly.

In a cubic crystal with ∇T along the x axis, phonon drag will give rise to an electric field

$$E_x = S_g(dT/dx) \tag{4.1}$$

This field exerts a force per unit volume of

$$F_x = n_0 e E_x \tag{4.2}$$

upon the electron gas of density n_0. Under steady-state conditions this must just equal the rate of momentum transfer to the electron gas from the phonons as a result of electron–phonon collisions. Assuming for the moment that the phonons interact with electrons only

$$F_x = \frac{d}{dt} \sum_q [N(q)\langle P_x(q)\rangle] = \sum_q \left[\left(\frac{\partial N(q)}{\partial t}\right)_{pe} \langle P_x(q)\rangle\right] \tag{4.3}$$

where $N(q)$ is the (nonequilibrium) number of phonons present, $\langle P_x(q)\rangle$ is the average x component of momentum transferred by a phonon of wave vector q to the electron gas, and $(\partial N(q)/\partial t)_{pe}$ is the rate of change of $N(q)$ due to the phonon–electron interactions.

The Boltzmann equation for phonons* is

$$\left(\frac{\partial N(q)}{\partial t}\right)_{collisions} = V(q) \cdot \nabla T\left(\frac{\partial N(q)}{\partial T}\right)$$

where $V(q)$ is the velocity associated with lattice vibrations of wave vector q. For a linear solution

$$N(q) = N_0(q) + N'(q)$$

where $N_0(q)$ is the equilibrium number of phonons and the Boltzmann equation for only phonon–electron interactions becomes

$$\left(\frac{\partial N(q)}{\partial t}\right)_{pe} = \left(\frac{\partial N'(q)}{\partial t}\right)_{pe} = V(q) \cdot \nabla T\left(\frac{\partial N_0(q)}{\partial T}\right)$$

Substituting into Eq. (4.3),

$$F_x = \sum_q V_x(q) \frac{dT}{dx} \frac{\partial N_0(q)}{\partial T} \langle P_x(q)\rangle \tag{4.4}$$

where the sum is over *all* q.

From (4.1) and (4.4) and using expressions for the conductivity and momentum associated with a free-electron gas,

$$\sigma = \frac{n_0 e^2 \tau}{m} \tag{4.5}$$

we obtain

$$S_g = \frac{e}{\sigma} \sum_q \frac{\partial N_0(q)}{\partial T} V_x(q) \langle \tau(v_x - v_{x'})\rangle \tag{4.6}$$

Here we have replaced $\langle P_x(q)\rangle$ by $\langle m(v_x - v_{x'})\rangle$, where v_x and $v_{x'}$ are the

---

*For example, see Equations (7.1.2) and (7.1.4) in Philip L. Taylor's *A Quantum Approach to the Solid State* [70T1].

velocities of the electron in the initial and final states, i.e., directly preceding and following scattering by a phonon of wave vector $\mathbf{q}$. For a cubic metal we finally arrive at the result

$$S_g = \frac{1}{3}\frac{e}{\sigma}\sum_{\mathbf{q}}\frac{\partial N_0(\mathbf{q})}{\partial T}\mathbf{V}(\mathbf{q})\cdot\langle\tau(\mathbf{v}(\mathbf{k})-\mathbf{v}(\mathbf{k}'))\rangle \tag{4.7}$$

It is this expression which we now compare with the sophisticated formal results of Bailyn.

According to Bailyn [67B1], S_g derived by the variational method is given by

$$S_g = \frac{\left|\frac{k}{e}\right|\frac{2}{3}e^2\sum_{\mathbf{q}j}\left\{\frac{\partial N_0(j\mathbf{q})}{\partial kT}\sum_{\mathbf{k}l,\mathbf{k}'l'}^{j\mathbf{q}}\alpha(j\mathbf{q};\mathbf{k}l,\mathbf{k}'l')[\mathbf{v}(\mathbf{k}l)-\mathbf{v}(\mathbf{k}'l')]\cdot\mathbf{V}(j\mathbf{q})\right\}}{\frac{2}{3}e^2\sum_{\mathbf{k}l}\mathbf{v}^2(\mathbf{k}l)(-\partial f_0/\partial\varepsilon)} \tag{4.8}$$

Here k is the Boltzmann constant, $\mathbf{v}(\mathbf{k}l)$ is the velocity of an electron with wave vector $\mathbf{k}$ on the lth sheet of the Fermi surface, and j indicates the polarization of the phonon of wave vector $\mathbf{q}$. A given $j\mathbf{q}$ phonon may be scattered by other phonons, by impurities, boundaries, etc., as well as by inducing electron transitions from states $\mathbf{k}l$ to $\mathbf{k}'l'$; in Eq. (4.8) the factor $\alpha(j\mathbf{q};\mathbf{k}l,\mathbf{k}'l')$ represents the probability that the phonon $j\mathbf{q}$ will induce the electron transition relative to all other scattering events in which this phonon could participate. Finally, the factor 2 in numerator and denominator comes from the sum over spin and the factor $\frac{1}{3}$ from replacing $V_x(j\mathbf{q})v_x(\mathbf{k}l)$ by $\frac{1}{3}\mathbf{V}(j\mathbf{q})\cdot v(\mathbf{k}l)$, valid for cubic materials.

We note immediately that since

$$\sigma = \frac{2}{3}e^2\sum_{\mathbf{k}l}\tau(\mathbf{k},l)v^2(\mathbf{k}l)(-\partial f_0/\partial\varepsilon) \tag{4.9}$$

we obtain

$$S_g = \frac{2}{3}\frac{e}{\sigma}\sum_{\mathbf{q}j}\frac{\partial N_0(j\mathbf{q})}{\partial T}\sum_{\mathbf{k}l,\mathbf{k}'l'}^{j\mathbf{q}}\alpha(j\mathbf{q};\mathbf{k}l,\mathbf{k}'l')[\tau v(\mathbf{k}l)-\tau v(\mathbf{k}'l')]\cdot\mathbf{V}(j\mathbf{q}) \tag{4.10}$$

provided τ is independent of $\mathbf{k}$. Evidently several points of similarity exist between the Bailyn result and our simplified expression Eq. (4.7). In both expressions the summation is over all $\mathbf{q}$, not just the nonequilibrium distribution. From the simple treatment it becomes apparent that the nonequilibrium nature is manifested through the $\partial N_0/\partial T$ factor, which, incidentally, decreases rapidly with $\mathbf{q}$. Also the origin of the $\mathbf{V}(j\mathbf{q})$ factor is clarified by the simple treatment. In Eq. (4.10) the factor $\langle P_x(\mathbf{q}) \rangle$ of Eq. (4.4) is replaced by

$$m \sum_{\mathbf{k}l,\mathbf{k}'l'}^{j\mathbf{q}} \alpha(j\mathbf{q}; \mathbf{k}l, \mathbf{k}'l')[\mathbf{v}(\mathbf{k}l) - \mathbf{v}(\mathbf{k}'l')] \qquad (4.11)$$

In this expression $j\mathbf{q}$ is constant during the summation over $\mathbf{k}, \mathbf{k}', l, l'$.

The Bailyn expression is obviously very complicated. To evaluate S_g from (4.10), one would need to know the phonon dispersion relationship, the transition probabilities for the $(j\mathbf{q}; \mathbf{k}l, \mathbf{k}'l')$ process and for all other phonon-scattering processes, and, finally, the electron velocities over the known Fermi surface. We shall see later that the above expression must be modified further to allow for effects arising from electron relaxation time anisotropies. First, however, we shall consider the Bailyn result with a view toward extracting from it some features that can be tested experimentally.

4.2 S_g at High Temperatures

Let us assume that α is independent of $j\mathbf{q}$; then we can write

$$\alpha = \tau_g/\tau_{pe} \qquad (4.12)$$

where τ_g is some mean phonon relaxation time, such as might enter into the theory of lattice thermal conductivity, and involves all phonon relaxation processes, and τ_{pe} is the phonon relaxation time for phonon–electron interactions only. At high temperature $\tau_g \sim \tau_{pp}$, the phonon relaxation time due to anharmonic phonon–phonon interaction. Since the temperature dependences of τ_{pp} and τ_{pe} are [58K1, 68B1]

$$\tau_{pp} \propto T^{-1} \qquad \tau_{pe} \propto T^0$$

it follows that, at high temperatures,

$$S_g \propto \alpha \propto T^{-1} \qquad (4.13)$$

Hence, in a rather idealized situation we might expect that the total thermoelectric power will exhibit a temperature variation of the form

$$S = AT + BT^{-1} \qquad (4.14)$$

where AT represents the theoretical diffusion term. Consequently, a plot of ST vs. T^2 should be a straight line whose intercept is B and whose slope is A.

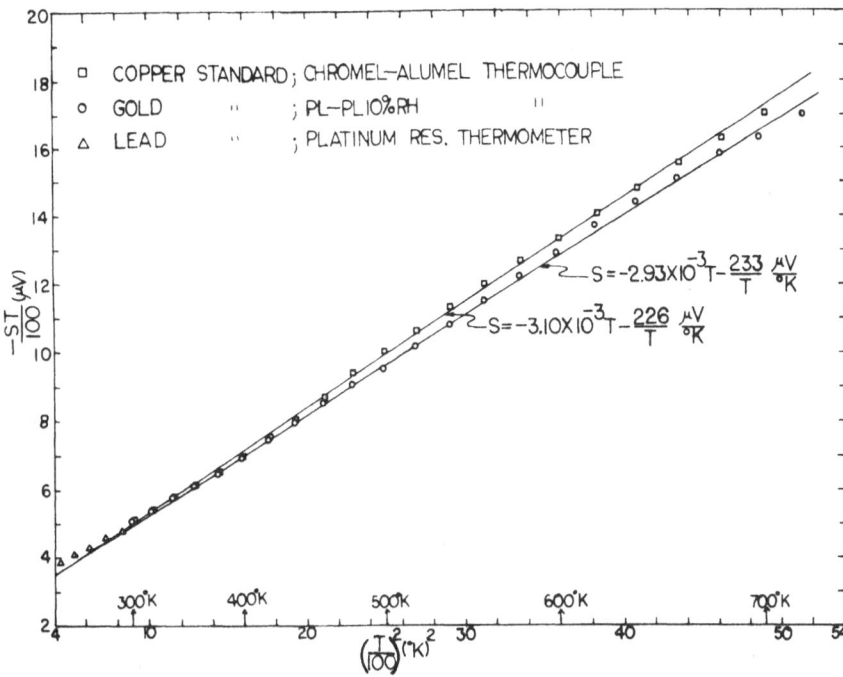

Fig. 4.1. A plot of ST vs. T^2 for pure aluminum over the temperature range 300 K to 700 K. The two different lines correspond to the use of copper and gold, respectively, as "standards" for the absolute thermopower at temperatures above 300 K. (See discussion in Chapter 2) (from [67G2]).

Generally one finds that even at temperatures well above the phonon-drag peak, which appears usually between $0.1\theta_D$ and $0.3\theta_D$, the graph of ST vs. T^2 is not a straight line. The results of Gripshover *et al.* [67G2] on aluminum (Fig. 4.1) appear to be the only ones that give good agreement with the simple theory over an appreciable temperature range, from about 300 to 700 K. In this metal, even at 700 K there appears to be still an appreciable phonon-drag contribution of about $0.2\,\mu V/K$ to the total thermopower of $2.4\,\mu V/K$. Powell [69P1] reports a linear plot of S vs. T^{-1} for gallium single crystals, but this is restricted to an interval of only 30 K; moreover, Powell ignores the possible presence of diffusion thermopower.

There are many possible reasons for the nonlinearity of ST vs. T^2 plots, i.e., for deviations from Eq. (4.14). First, the plot presumes that the diffusion thermopower varies linearly with temperature. Although this is consistent with elementary theory, experimental support for this is not strong and is limited to relatively narrow temperature regions. Second, turning our attention to S_g, every sum over $\mathbf{k}, \mathbf{k'}, l, l'$ in Eq. (4.10) is multiplied by $\partial N_0(j\mathbf{q})/\partial T$, which, for fixed $\mathbf{q}$, is strongly temperature dependent. Third, dispersion of the phonon spectrum for large values of $\mathbf{q}$ will tend to reduce S_g with increasing temperature [64B1]. All in all, the surprising thing is not the apparent departure of the phonon-drag component from a T^{-1} dependence at high temperatures, but that aluminum should exhibit this dependence so closely.

4.3 S_g at Low Temperatures

As we saw in Chapter 2, simple arguments suggest that at low temperatures S_g should be proportional to the lattice heat capacity C_g, and, therefore, proportional to T^3. If we assume a spherical Fermi surface and invoke the Debye approximation, this result also follows directly from the Bailyn equation.

With the above assumptions we replace $\mathbf{v}(\mathbf{k})$ by $\hbar \mathbf{k}/m$ and $\mathbf{V}(\mathbf{q})$ by $\mathbf{q}\omega(q)/q^2$ in Eq. (4.10); we also set $\alpha(j\mathbf{q}; \mathbf{k}l, \mathbf{k'}l') = 1$, which is valid for a pure metal at low temperatures. If we now assume that only N processes need be considered so that

$$\mathbf{v}(\mathbf{k}) - \mathbf{v}(\mathbf{k'}) = \frac{\hbar}{m}(\mathbf{k} - \mathbf{k'}) = -\frac{\hbar}{m}\mathbf{q}$$

we obtain

$$S_g = -\frac{2|e|}{3\sigma}\sum_q (\partial N_0/\partial T)(\hbar q/m)[\omega(q)\tau/q]$$

$$= \frac{2}{3n_o e}\sum_q (\partial N_0/\partial T)\hbar\omega(q) = \frac{2}{3n_o e}C_g = bT^3 \qquad (4.15)$$

If in this range the diffusion thermopower is proportional to T, we expect that the total thermopower will be given by

$$S = aT + bT^3 \qquad (4.16)$$

Consequently, a plot of S/T vs. T^2 should yield a straight line whose slope gives the phonon-drag coefficient b and whose intercept gives the diffusion coefficient a. Before discussing the experimental results, we shall consider briefly the validity of the above assumptions. First, as regards the Debye model this is probably a reasonable assumption at sufficiently low temperatures, 4.2 K or less. Second, the use of a free-electron gas model is surely inappropriate for all except the alkali metals, and we shall indeed consider the experimental information for these metals in some detail. The complex Fermi surfaces of all other metals will allow for a variety of phonon–electron scattering events which, as we pointed out in Chapter 2, can give rise to positive or negative contributions and have a temperature dependence that deviates substantially from that of the lattice specific heat. Third, neglect of Umklapp scattering is a serious limitation of this simple theory. We discussed the role of U processes in Chapter 2 and found that in the free-electron gas model they should contribute a positive term to S_g, canceling in part or overcompensating for the negative contribution arising from N scattering. In the alkali metals the situation depicted in Fig. 2.7 should apply, with the result that at low temperatures S_g^U would decay exponentially.

We recall that if the Fermi surface does not touch a zone boundary, only phonons with wave vector larger than $\mathbf{q}_{min}$ can induce U processes. Assuming a Debye distribution, the number of phonons of wave vector $\mathbf{q}_{min}$ is

$$N_0(\mathbf{q}_{min}) \propto [\exp{(\hbar v q_{min}/kT)} - 1]^{-1} \tag{4.17}$$

where v is the velocity of sound. If $\mathbf{q}_{min}$ is sufficiently large so that at the temperature of interest $\exp{(\hbar v q_{min}/kT)} \gg 1$, we have

$$N_0(\mathbf{q}_{min}) \propto \exp{(-\theta^*/T)} \tag{4.18}$$

where $\theta^* = \hbar v q_{min}/k$. Consequently, we expect that under these circumstances the number of U events will diminish exponentially with decreasing temperature. At very low temperatures (and this depends, of course, on the value of θ^*) S_g^U would become negligibly small compared to the normal phonon-drag and diffusion contributions. When this situation prevails, it may then be possible to observe the theoretical T^3 variation of the normal phonon-drag contribution. For metals other than the alkalis the Fermi surface either touches a zone boundary or is multisheet. In these cases Umklapps will always be a complicating factor down to the lowest temperatures.

4.3a *Low-Temperature Phonon Drag in the Alkali Metals*

The Fermi surfaces of some of the alkali metals have been examined very precisely by the de Haas–van Alphen techniques. Unfortunately lithium and sodium undergo partial phase transformations at low temperatures: from the body-centered cubic structure at room temperature to

Table 4.1.

Element	Lattice parameter $(10^{-8}\,\mathrm{cm})$	K_{110} $(10^7\,\mathrm{cm}^{-1})$	$q_{\min} = 2[(\pi\sqrt{2}/a) - K_{110}]$ $(10^7\,\mathrm{cm}^{-1})$
Li	3.5	11.4	2.54
Na	4.28	9.2	2.36
K	5.33	7.5	1.66
Rb	5.62	7.0	1.60
Cs	6.05	6.6	1.58

hexagonal close packed for sodium, and to hexagonal close packed and face-centered cubic for lithium. This makes lithium and sodium unsuitable candidates for detailed examination. Of the remaining alkalis, potassium has a very nearly spherical Fermi surface while rubidium and cesium have Fermi surfaces that are progressively more distorted. The Fermi surface is pulled toward the zone boundary resulting in smaller values of $q_{\min}$ as indicated in Table 4.1. From these values we anticipate that the effect of Umklapps would decrease most rapidly with temperature for potassium, and that this is the most likely element in which to find the normal phonon-drag T^3 dependence at presently measurable temperatures.

The thermoelectric properties of the alkali metals have been studied extensively by the group at the National Research Council of Canada at Ottawa [58M1, 58M2, 58M3, 60M1, 61G1, 63G1].

For their lowest temperature measurements, the hot end of their specimens was maintained at the constant temperature of a pumped liquid ^{4}He bath. The other end was thermally connected to a "pill" of a paramagnetic salt which was cooled to about 0.1 K by adiabatic demagnetization. The thermo-emf vs. a superconductor was then measured as the cold junction warmed, and the thermopower obtained by differentiation of the e.m.f.–temperature curve.

Initially, the Ottawa group fitted the data below 3 K for their "purest" alkalis to the expression

$$S = AT + BT^3 + C e^{-\theta^*/T} \tag{4.19}$$

or alternatively

$$E = \tfrac{1}{2}AT^2 + \tfrac{1}{4}BT^4 + D e^{-\theta^*/T} \tag{4.20}$$

The results are outlined in Table 4.2.

Table 4.2.

Metal	Approximate residual resistance ratio, $R_{4.2}/R_{300}$	A (nV/deg^2)	B (nV/deg^4)	C (nV/deg)	D (nV)	θ^* (K)
Li	1.3×10^{-3} to 2.5×10^{-3}	–	–	–	–	–
Na	2.3×10^{-4} to 6.9×10^{-4}	-0.5 to -30	-0.2 to -0.8	–	$+100$ to $+200$	8 to 20
K	9.6×10^{-4} to 2.7×10^{-3}	$+0.5$ to -10	-1.5 to -3.0	–	$+480$	~21
Rb	2.8×10^{-3} to 30×10^{-3}	$+20$ to -2	0 to -6	5×10^3 to 5×10^4	5000	~16
Cs	1.4×10^{-3} to 16×10^{-3}	$+50$	-23	8×10^{-3}	–	~6

From these preliminary data, it would seem that the interpretation we have given so far is at least qualitatively correct. θ^* for sodium and potassium is so high that the last term in Eqs. (4.19) and (4.20) could be ignored at very low temperatures. As we move from sodium to cesium, θ^* becomes progressively smaller, meaning that measurements to be independent of the Umklapps must be made at lower and lower temperatures. The results for sodium seemed to be remarkably independent of the martensitic transformations. The variation of A from sample to sample is an indication that the diffusion thermopower at these low temperatures is being controlled by poorly characterized impurity scattering rather than by phonon-scattering characteristics of the pure metal.

With this work as background, Guénault and MacDonald performed more detailed measurements on potassium alloys between ~0.1 and 3 K. On the basis of the large θ^* obtained above for potassium, they completely ignored the Umklapp contribution to the phonon-drag thermopower. Further evidence of the high value of θ^* for potassium comes from the resistivity studies of Gugan [71G1] and of Ekin and Maxfield [71E2]. Gugan obtains a value of θ^* of about 23 K but comments that the resistivity data are not a crucial test of the exponential falloff of the Umklapp contribution. Ekin and Maxfield also find a value of about 20 K and analyze some of the finer points of the theory in some detail.

For the normal contribution Guénault and MacDonald use the theoretical expression [56K2, 56S1] valid for free electrons and the Debye approximation

$$S_g^N = 3\frac{k}{e}\left(\frac{T}{\theta_D}\right)^3 \int_0^{\theta_D/T} \frac{X^4 e^{-X}}{(1-e^{-X})^2} f(X)\, dX \qquad (4.21)$$

where $X = \hbar\omega/kT$, and θ_D is the Debye temperature. Here $f(X)$ is the fractional probability of a phonon being scattered by an electron rather than by anything else, i.e., $f(X)$ is the equivalent of α of Eq. (4.10). Equation (4.21) is entirely equivalent to Eq. (4.15) derived from the Bailyn equation when $f(X) = 1$.

In terms of phonon mean free paths,

$$f = \frac{1/\Lambda_e}{1/\Lambda_r + 1/\Lambda_e} \qquad (4.22)$$

where Λ_r and Λ_e are phonon mean free paths for impurity and electron scattering, respectively; phonon–phonon scattering is neglected at these low temperatures. According to Klemens [56K2], for point defect scattering

$$\Lambda_r = A\omega^{-4}$$
$$\Lambda_e = B\omega^{-1} \tag{4.23}$$

where A and B are constants. Hence,

$$f = \frac{1}{1 + C\omega^3} \tag{4.24}$$

Defining a "scattering temperature" ϕ by

$$B/A = C = \left(\frac{\hbar}{k\phi}\right)^3 \tag{4.25}$$

we now have

$$f = \frac{1}{1 + (T/\phi)^3 X^3} \tag{4.26}$$

High values of ϕ correspond to low impurity concentrations and vice versa.

To evaluate S_g, f from Eq. (4.26) is inserted in Eq. (4.21) and the integration is then performed numerically for a range of values of ϕ using $\theta_D = 100$ K for potassium. The results are in semiquantitative agreement with the data. They show a T^3 dependence for $T \ll \phi$ when $f \approx 1$ and become constant for $T \gtrsim \phi$. They also have the correct concentration dependence built in. For high concentration ϕ is small and the phonon-drag thermopower rapidly saturates. Guénault and MacDonald developed the theory one stage further by allowing transverse and longitudinal phonons different values of θ_D ($\theta_L = 130$ K, $\theta_T = 90$ K) and of ϕ. Then

$$S_g^N = \tfrac{1}{3} S_g(\theta_L, \phi_L) + \tfrac{2}{3} S_g^N(\theta_T, \phi_T) \tag{4.27}$$

In the analysis of their data, they assumed that below 0.5 K the thermopower was entirely diffusion thermopower. The diffusion thermopower at higher temperature was obtained by assuming that S_d is proportional to temperature. Deviations of the measured thermopower from the diffusion thermopower were then fitted to Eq. (4.27) by choosing the most favorable values of ϕ_L and ϕ_T. The results are shown in Table 4.3 for potassium–rubidium alloys. (Data for alloys with sodium and cesium are given in the original paper.) The experiments indicate that $\phi_L \sim 10\phi_T$. If we assume that in (4.23) $A_L = A_T$, i.e., the impurity mean free path is the same for

Table 4.3.

Rb (at.%)	$\dfrac{10^3 R_{4.2\,K}}{R_{293\,K} - R_{4.2\,K}}$	S_d at 3 K $(10^{-8}\,\text{V/deg})$	ϕ_L (K)	ϕ_T (K)
0.2	6.4	+0.65	50	6.5
0.9	22	+1.45	45	4.5
2.1	39	+1.9	20	2.5
12	218	+2.0	3.5	<1
24	339	+2.2	~2	<1

transverse and longitudinal waves, then from the definition of ϕ

$$\left(\frac{\phi_T}{\phi_L}\right)^3 = \frac{B_L}{B_T}$$

which means that the ratio of the phonon–electron interaction constants is

$$\frac{B_L}{B_T} \sim \frac{1}{1000}$$

This is consistent with the known small interaction of transverse lattice vibrations with electrons.

For potassium, then, the general features of the thermopower appear to be well understood. Detailed calculations involving phonon–electron and phonon–impurity interactions are yet to be made. As the theory is refined, it may become desirable to extend the measurements to lower temperatures to define the diffusion thermopower more precisely.

The study of rubidium and cesium alloys at very low temperatures is complicated by the fact that θ^* is much smaller in these elements. The preliminary results of MacDonald *et al.* [60M1] indicate values of 16 K and 6 K for rubidium and cesium, and, corresponding with these lower values, Umklapp processes occur significantly at 1.2 K for cesium and 1.5 K for rubidium. It is therefore necessary to derive S_d by examination of the very low-temperature limit of the experimental curves. Thereafter, it is assumed that the difference between the measured and diffusion thermopower is entirely Umklapp phonon-drag thermopower, and the expression

$$S_g = A\,e^{-\theta^*/T} \tag{4.28}$$

is tested by plotting $\ln S_g$ vs. $1/T$. Good straight lines are obtained between ~1.2 and 3 K, with values of θ^* the same order of magnitude as those noted above. The value of θ^* is, of course, related to the distortion of the Fermi surface from a sphere. If we recall the definitions

$$\theta^* = \frac{\hbar v q_{min}}{k} \tag{4.29}$$

$$\theta_D = \frac{\hbar v q_{max}}{k} \tag{4.30}$$

we have

$$\frac{\theta^*}{\theta_D} = \frac{q_{min}}{q_{max}} \tag{4.31}$$

For an undistorted one-electron sphere we obtain

$$\frac{\theta^*}{\theta_D} = 0.19$$

and progressive distortion will reduce this to zero when the Fermi surface contacts the zone boundary. Experimental values $\theta^*/\theta_D = 0.22, 0.24, 0.14$ for potassium, rubidium, and cesium, respectively, are consistent with the Fermi surface data for the alkali metals which indicate that the Fermi surfaces of potassium and rubidium are nearly spherical but that there are considerably greater distortions in cesium.

4.3b Low-Temperature Phonon Drag in Other Metals

We have already commented that for all metals other than the alkali metals the Fermi surface touches one or more zone boundaries. The minimum phonon wave vector for Umklapps to occur is then zero, and Umklapps must be considered down to the lowest temperatures. A priori, one would not anticipate a T^3 law to be obeyed at low temperatures. Despite this, many researchers have analyzed their results in terms of the simple expression

$$S = aT + bT^3 \tag{4.16}$$

The first essential requirement for the application of (4.16) is that the metal be in the domain where the diffusion thermopower is expected to be proportional to T, i.e., in the residual resistance region. In general,

$$S_d(T) = \frac{W_r}{W_r + W_i} S_d^r + \frac{W_i}{W_r + W_i} S_d^i \tag{2.46}$$

S_d^r but not S_d^i is expected to be proportional to T at low temperature.* Hence, we expect linearity of $S_d(T)$ only when $W_i S_d^i \ll W_r S_d^r$. The purest metals may therefore not be the best materials to investigate, unless the phonon-drag component is much larger than the diffusion thermopower.

We shall now consider the evidence from measurements on various metals. As regards the noble metals, generally copper and gold have not been good candidates because of the effect of dissolved iron (see Chapter 6). This effect can be reduced by oxidation, and Guénault [74G5] has performed satisfactory low-temperature measurements on homovalent copper alloys in bulk form in oxidized condition. Silver, on the other hand, frequently does not show a Kondo minimum, not because it is free of iron but because normally it is oxidized. We shall therefore consider the measurements on silver in detail.

Pearson's [60P1] early measurements on silver with a resistance ratio $R_{294}/R_{4.2} = 364$ suggest a roughly T^2 dependence of the phonon-drag thermopower in the range 3–17 K. This is for a sample for which $(W_i/W_r) \sim$ 0.1 at 10 K so that impurity scattering should have dominated the diffusion thermopower contribution. Van Baarle et al. [66V1], using purer samples with $\rho_{273}/\rho_{4.2}$ as high as 1800 and for which $W_r \sim W_i$ at 10 K, could only conclude that "a T^3 relation is possibly valid at the lowest temperatures (2 K) used in these experiments." Similarly, results by Van Baarle et al. [67V1] on Ag–Sb, Ag–Au, and Au–Pd alloys do not rigorously verify the T^3 law.

Perhaps the most detailed work on a noble metal was performed by Guénault [67G1] on silver and silver alloys. Guénault used a helium-3 cryostat to permit measurements in the range between 0.3 and 5 K. The theoretical form of the low-temperature thermopower he used was

$$S = aT + bT^3 + CT/(T + T_0) \qquad (4.32)$$

where the third term is a correction to the diffusion term arising from traces of magnetic impurity. It has the form suggested by Kondo [65K1]. The integral of the thermopower

$$E = \frac{1}{2} aT^2 + \frac{1}{4} bT^4 + CT \left[1 - \frac{T_0}{T} \log \left(1 + \frac{T}{T_0} \right) \right]$$

was directly compared with experimental results for the thermo-emf. This was found to be "adequate" to fit the results up to 4 K and often to 5 K. By

---

* But see P. G. Klemens's article in *Physica* [73K3] for a discussion of this point.

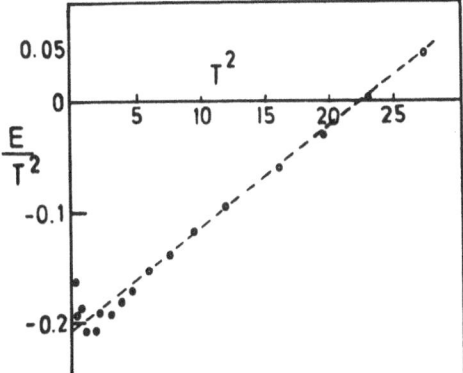

Fig. 4.2. The absolute thermoelectric voltage of an Ag + 1% In alloy at low temperatures (from [67G1]).

"adequate" one would presume that the data fit an expression of this form within the experimental error. For those samples in which the iron content was so low that C was negligibly small, E/T^2 could be plotted against T^2 to obtain a straight-line graph, as shown in Fig. 4.2. Between 2 and 5 K for this Ag + 1% In sample the scatter in the points is sufficiently small to clearly distinguish a T^3 from a T^2 or T^4 dependence of the phonon-drag thermopower. Below 2 K there is considerable scatter and the results are not so conclusive. For future reference we give some of Guénault's results in Table 4.4. We note that a and b are of the same order of magnitude, and this results in the phonon-drag and the diffusion thermopowers also being of the same order of magnitude. Note that the addition of Ge, Tl, In, or Pd has very little effect on the value of b, whereas the addition of Au has a very pronounced effect. We shall return to this point in Section 4.4.

Later measurements by Pemberton and Guénault [71P1] on dilute copper alloys give much larger values of C and smaller values of b, and as a consequence considerably larger standard errors. The measurements cannot, therefore, be taken as strong evidence of a T^3 law in copper. In his work on homovalent copper alloys Guénault [74G5] succeeded in largely eliminating the iron by oxidation and thereby obtained improved values of b. An example of his data for a 1.9 at.% alloy of gold in copper is given in Table 4.4. It is clear that b is an order of magnitude lower than that in the comparable silver alloy. The phonon-drag term is considerably less than the diffusion term over the whole range of measurement. Again the results for the silver alloys must be taken as much stronger evidence of the T^3 law.

Table 4.4.

Solute	Nominal at.%	Resistivity, approximate ($\mu\Omega$-cm)	a 10^{-9} V/K^2	b 10^{-9} V/K^4	c 10^{-9} V/K
Silver Alloys					
None	2 ppm O$_2$	0.014	+2.4	0.36	<0.2
Ge	1	4.78	−2.3	0.31	0
	0.5	2.52	−2.6	0.34	0.5
	0.2	1.00	−2.2	0.32	3.8
Tl	1	1.79	−3.9	0.36	0
	0.5	0.95	−3.8	0.38	3.3
	0.25	0.46	−3.8	0.38	7.6
In	2	3.10	−3.2	0.32	0
	1	1.65	−4.1	0.36	0
	0.5	0.82	−4.1	0.34	2.8
	0.2	0.34	−4.0	0.38	<0.5
Au	2	0.64	−3.5	0.68	1.4
	1	0.36	−2.8	0.72	1.4
	0.5	0.19	−3.3	0.70	8.5
Copper–Gold Alloys					
Cu–Au vacuum annealed	1.9	0.92	2.8	0.043	−72
Cu–Au oxidized	1.9	0.92	4.4	0.026	−22

Worobey *et al.* [65W2] performed measurements on oxidized gold films, which were prepared by evaporating 99.999% gold onto a glass substrate at a pressure of 4×10^{-7} Torr. The films were then heated in air at 450°C. The data were fitted to the form

$$S = A + aT + bT^3 \qquad (4.33)$$

where a and b have their previous significance, and A is presumed to account for the contribution from ferromagnetic impurities. The results in Fig. 4.3 show that after heating for 85 hours, $A \to 0$ and the plot of S/T vs. T^2 is a good straight line for 1.4 K $< T <$ 8 K with $a \sim 0.32 \times 10^{-8}$ V/K^2 and $b \sim 1.3 \times 10^{-10}$ V/K^4. Consequently, over most of the temperature range considered the phonon-drag thermopower is small compared with the diffusion thermopower. The S/T vs. T^x plot becomes rather insensitive to x under these circumstances.

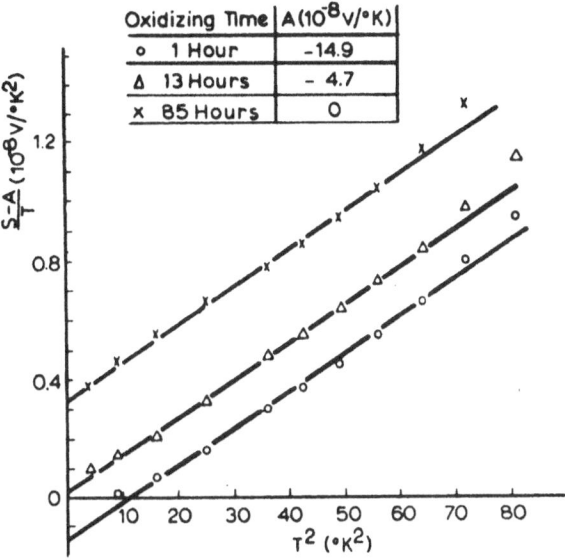

Fig. 4.3. The thermoelectric power of gold films subject to annealing at 450°C for various times. The curves are fits to Eq. (4.33). Values of the parameter A are shown in the insert (from [65W2]).

Aluminum has been studied by several groups commencing with De Vroomen *et al.* [60D1] in 1960. For pure aluminum the plot of S/T vs. T^2 was linear from 2 to 5 or 6 K. For alloys containing 0.8 and 1.2% Mg, the region extended from 2 to 8 K. We have listed values of b in Table 4.5 along with the results of subsequent workers. The addition of Mg makes no appreciable difference to b. Boato and Vig [67B5] performed similar measurements on aluminum containing transition metal solutes (Ti, V, Cr,

Table 4.5.

Authors	Material	b (10^{-10} V/K^4)
De Vroomen *et al.* [60D1]	Al, Al+Mg	−1.4
Boato and Vig [67B5]	Al+Ti, V, Cr, Mn, Fe	−1.85
Averback *et al.* [73A1]	Highest RRR Al	−0.7
	Al+Sn	−0.8
	Al+Cu	−1.7
	Al+Cd	−1.2
	Lower RRR Al	~−1.3

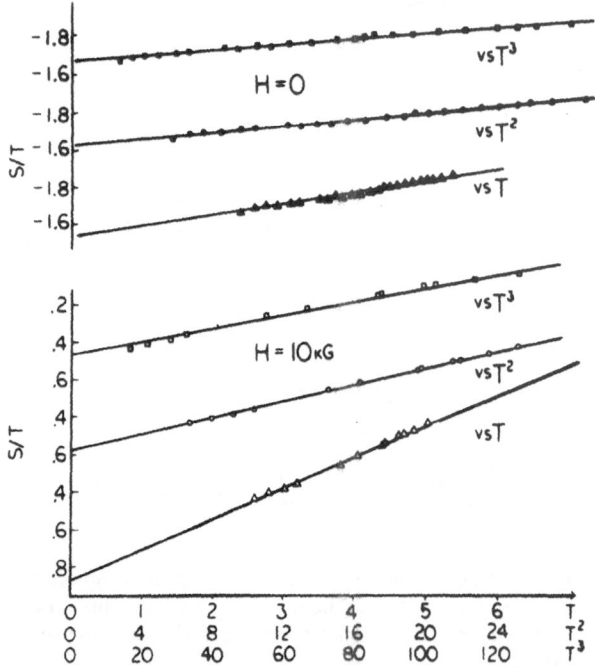

Fig. 4.4. A plot of S/T vs. T^n for an aluminum + 100 ppm tin alloy
for $n = 1, 2,$ and 3. Results are shown for zero magnetic field and
for a field of 10 kG (from [73A1]).

Mn, and Fe) and obtained a T^3 region from ~1.5 to 6 K. Again b was
independent of the added impurity.

Averback, Stephan, and Bass [71A1, 73A1] have recently performed
measurements on aluminum and aluminum alloys over the range 2.0 to 5 K,
both in and out of a magnetic field, and have analyzed their results using Eq.
(4.16). Resistance ratios ranged from 4600 for the best Al to 97 for an
Al–Cu alloy. In all instances the graphs of S/T vs. T^2 were good straight
lines. However, in contrast to silver, in aluminum $b \ll a$, and under these
circumstances distinguishing the T^3 from a T^2 or a T^4 law is not so conclusive
as illustrated by their results on Al–Sn alloy RRR = 1200 shown in Fig. 4.4.
We note that the scatter in their points was considerably less than those of
De Vroomen *et al.* and of Boato and Vig, and, consequently, all the
aluminum data must be treated with some caution. For future reference we
note (see Table 4.5) that the addition of Cu and perhaps Cd to Al tends to
increase the magnitude of the phonon-drag thermopower.

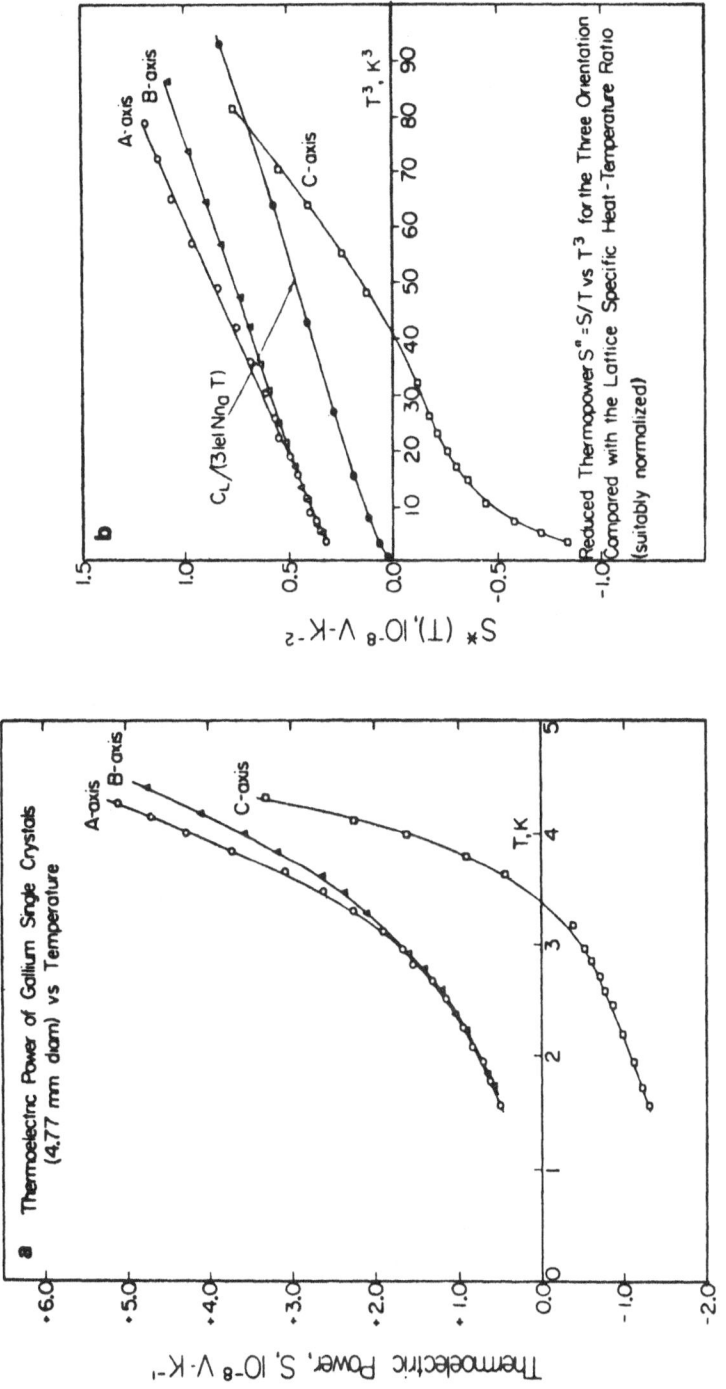

Fig. 4.5. The thermoelectric power of single-crystal gallium. (a) S vs. T and (b) S/T vs. T^3 along A, B, and C axes. (from [73M1]).

Gold and Pearson [61G3] find a linear plot for S/T vs. T^2 for lead, but only in the region between 7 and 9 K. The remaining simple (nontransition or rare-earth) metals are not cubic, and for them the components of the thermoelectric power tensor must be measured on single crystals to be of much significance. Rowe and Schroeder's [70R1] measurements on Mg, Cd, and Zn (see Fig. 2.5) certainly do not show any evidence of a T^3 law—nor do those of Caplin *et al.* [74C2] on indium. In Ga, a T^4 dependence appears to be obeyed along the A and B axes but not along the C axis [73M1] as shown in Fig. 4.5.

On the basis of the available evidence one is forced to conclude that the experimental evidence for a low-temperature T^3 law for the phonon-drag thermopower in real metals is very meager. The most positive results are those of Guénault on silver-based alloys. Therefore one should be very wary in the use of Eq. (4.16); the theory on which it is based can be expected to break down at many points. First, it is a free-electron theory which ignores Umklapps. Second, it relies on the Debye approximation and it is well known that the Debye temperature fluctuates wildly in the low-temperature region where one would expect the T^3 law to be obeyed [55B1]. A further complication which limits the upper bound of the region where the T^3 rule is expected is that $\alpha(q)$ falls below 1 because of phonon–phonon interaction. The peak of the phonon-drag thermopower after which phonon–phonon scattering dominates over phonon–electron scattering is typically 0.1 to 0.3 θ_D; the upper bound of the T^3 region must be considerably less than this. At temperatures in the dilution refrigerator and He^3 cryostat range, the T^3 law may be obeyed, but precision experiments to test this have yet to be performed.

4.4 Anisotropy of Relaxation Times and Phonon-Drag Thermopower

When impurities are added to a metal, one of the obvious effects on the phonon-drag thermopower is that, because of phonon–impurity scattering, $\alpha(\mathbf{q})$ will become less than 1 even at low temperatures. Consequently, one would expect the phonon-drag thermopower to decrease with increasing impurity concentration. This is frequently the case, and the effects at higher temperatures based on this concept will be examined in Section 4.6. There are notable exceptions, however, especially at low temperatures. For example, the results of Van Baarle [67V2] on some silver-based alloys shown

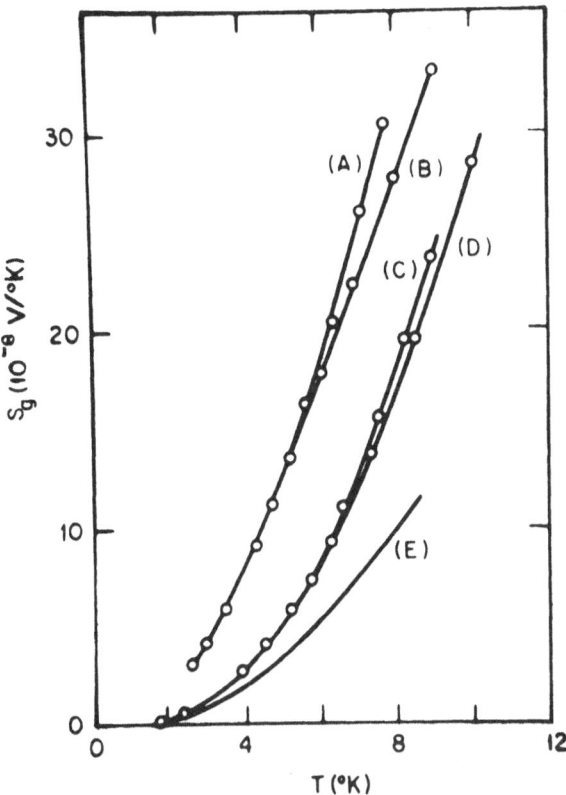

Fig. 4.6. S_g of some dilute silver alloys: (A) 0.3% Au, (B) 1.0% Au, (C) 0.3% Sb, (D) 0.19% Sb. (E) is the estimated phonon-drag thermopower of pure silver (from [67V2]).

in Fig. 4.6 indicate that the addition of gold and antimony to silver results in an increased phonon-drag thermopower. To obtain S_g the raw data were fitted to Eq. (4.16) and S_d was subtracted from the measured thermopower. Similarly, we have seen from Guénault's results that the addition of gold to silver doubles the value of b. Less definite effects for aluminum have been mentioned above (Table 4.5). According to Huebener's results [68H2] the magnitude of the phonon-drag thermopower increases below ~ 45 K for dilute alloys with zinc and silver in aluminum, and increases below ~ 75 K for dilute alloys of magnesium in aluminum.

As described above, α in Eq. (4.8) may vary with impurity content but will always act to decrease the magnitude of the phonon-drag thermopower. To produce a possible increase in thermopower, the Bailyn equation has to be modified. There is an effect due to the impurity scattering of electrons, but it has been lost in the approximation of Bailyn's variational approach. To retrieve it, forsaking the variational methods, Bailyn [67B1] solved the generalized Boltzmann equation by assuming a relaxation time $\tau(\mathbf{k}, l)$ for electron–impurity scattering and a phonon relaxation time for all phonon interactions. The result is very similar to Eq. (4.8):

$$S_g = \frac{|k/e|\frac{2}{3}e^2 \sum_{jq}[\partial N_0(\mathbf{q}, j)/\partial kT]\sum_{\mathbf{k}l\mathbf{k}'l'}^{jq}\alpha(jq; \mathbf{k}l, \mathbf{k}'l') \cdot Z}{\frac{2}{3}e^2 \sum_{\mathbf{k}l} v(\mathbf{k}l)^2\tau(\mathbf{k}l)(-\partial f_0/\partial\varepsilon)} \tag{4.34}$$

where

$$Z = [\mathbf{v}(\mathbf{k}l)\tau(\mathbf{k}l) - \mathbf{v}(\mathbf{k}'l')\tau(\mathbf{k}'l')] \cdot \mathbf{V}(j\mathbf{q})$$

and becomes identical with Eq. (4.8), which does not contain τ, when τ is a function of energy only and not of $\mathbf{k}$ or l. At low temperatures where electron–impurity scattering predominates, this relaxation time equation should make a reasonable approximation.

The influence of electron–impurity scattering is now manifested through the relaxation time $\tau(\mathbf{k}, l)$. To appreciate the role of this relaxation time, particularly that of its anisotropy in the present instance, we divide the Fermi surface into different sections, labeled i, for each of which the electrons display essentially constant properties (relaxation time, effective mass, and Fermi velocity). The conductivity of each of these regions is given by

$$\sigma_i = \frac{2}{3}e^2 \sum_{\mathbf{k}l}^{(i)} \tau(\mathbf{k}l)[v(\mathbf{k}l)]^2(-\partial f_0/\partial\varepsilon) \tag{4.35}$$

The phonon-drag thermopower associated with this region is denoted by S_{gi}, equal to an expression identical with Eq. (4.34) except that the second summation in the numerator and the summation in the denominator are over the ith region only. If the temperature is sufficiently low, the $\mathbf{q}$'s are in general much smaller than the dimensions of each region so that interactions involving phonons between adjacent regions can be neglected. It then follows that

$$S_{gi} = \left|\frac{k}{e}\right|\frac{2}{3} e \frac{\sum_{jq}\sum_{\mathbf{k}l\mathbf{k}'l'}^{i}}{\sigma_i} \tag{4.36}$$

and

$$S_g = \left|\frac{k}{e}\right|\frac{2}{3} \frac{\sum_{jq}\sum_{\mathbf{k}l\mathbf{k}'l'}}{\sigma} = \sum_i \frac{\sigma_i}{\sigma}S_{gi} \tag{4.37}$$

Here we have abbreviated the notation by omitting the quantities appearing in the summations over $j\mathbf{q}$ and $\mathbf{k}l\mathbf{k}'l'$.

A similar expression can be derived when T is so high that interregion transitions cannot be neglected, but the expression for S_{gi} is rather more complex.

Since the relaxation times have been assumed uniform over each of the regions, they cancel out of each of the S_{gi}'s. The sign of each contribution to the thermopower is then determined by whether Z is positive or negative.

As was shown in Chapter 2 (page 36), for simple closed Fermi surfaces the contribution to the phonon-drag thermopower is positive or negative depending on whether the $\mathbf{q}$ vector connects concave or convex regions of the Fermi surface.

We now consider how the above equations lead to a consistent interpretation of the experimental data as far as alloys of the noble metals are concerned, and for this we follow the approach of Dugdale and Bailyn [67D1]. The Fermi surface of the noble metals (see Fig. 2.4) is divided into three regions, i.e., the subscript i takes on values 1 to 3. Region 1 is the convex belly region associated with $\langle 100 \rangle$ directions. The electrons in this region are predominantly s-like in character and have little d-admixture, although there are significant departures from free-electron character for gold. Region 2 is the concave belly region centered around $\langle 110 \rangle$ directions and is associated with appreciable d-admixture [61S1, 62R1, 61M1]. Region 3 is the neck region, which is concave in the $\langle 111 \rangle$ direction and convex in the (111) plane. Here the electron wave functions have a large p-component. Using the subscripts cv, cc, and n to denote these regions, Eq. (4.37) becomes

$$S_g = \frac{\sigma_{cv}}{\sigma} S_{g,cv} + \frac{\sigma_{cc}}{\sigma} S_{g,cc} + \frac{\sigma_n}{\sigma} S_{g,n}$$

We assume τ is constant within each region so that it cancels out of each of the S_{gi}. The S_{gi} then involve phonon interactions only through the α terms of Eq. (4.34). Furthermore, the dominant *phonon* interaction at low temperatures (~ 1 K) is phonon–electron scattering rather than phonon–impurity scattering. Hence, even though the *electron* relaxation times are determined by electron–impurity scattering, the S_{gi} involve effects due to electron–phonon interaction only and depend exclusively on the properties of the host material. The effect of adding different types of impurities is contained entirely in σ_i/σ, which will be independent of electron–phonon scattering at the low temperatures considered here. Thus the problem neatly resolves itself into two parts. One part involves S_{gi} and electron–phonon scattering, and the other σ_i/σ and electron–impurity scattering. Detailed consideration of electron and phonon properties in these metals, including anisotropies of the Fermi and phonon velocities, leads to the following conclusions. Signs: $S_{g,cc}$ positive; $S_{g,cv}$ negative; $S_{g,n}$ contributions of both signs. Positive contribution dominates. We find

for copper: $S_{g,n} \sim 3S_{g,cc} \sim 30S_{g,cv}$

for silver: $S_{g,n} \sim 10S_{g,cc} \sim 100S_{g,cv}$

for gold: $S_{g,n} \sim 2S_{g,cc} \sim 6S_{g,cv}$

In estimating the role of various impurities on σ_i, i.e., on τ_i, Bailyn and Dugdale divide impurities into three groups as follows: (A) Well-localized uncharged impurities. To these they ascribe an s-like character; that is, there are large s-phase shifts for scattering at η. (B) Charged impurities with roughly equal s- and p-phase shifts and smaller but not negligible d-shifts. (C) Transition metal impurities which have a strong d-character.

They then make extensive use of the following rule of thumb: large scattering occurs when an impurity of certain character interacts with electrons whose wave functions are of the same character. Large scattering from region i means small τ_i and a small contribution from the σ_i/σ factor. The following conclusions emerge.

1. An uncharged impurity (type A) interacts mainly with the s-waves associated with the convex part of the Fermi surface. This tends to decrease σ_{cv}, and, consequently, the magnitude of negative contributions of the convex parts to S_g.

2. A charged impurity (type B) will scatter more in s- and p-like regions associated with convex belly and neck electrons, and less in the d-like concave regions. Consequently, $\tau_{cc} \gg \tau_{cv}$ or τ_n, and the positive contribution to S_g from the concave regions is enhanced.

3. A transition metal (type C) impurity will favor interactions with the concave regions giving a small τ_{cc} and will tend to decrease the positive concave term.

These conclusions agree well with the relaxation times over the Fermi surface found from the de Haas–Van Alphen effect [74P3].

As an example, we consider the application of these concepts to silver, for which we have seen $S_{g,n} \gg S_{g,cc}$ or $S_{g,cv}$. From Matthiessen's rule experiments [67D2], $\sigma_n/\sigma \sim 1$. Consequently,

$$S_g \sim \frac{\sigma_n}{\sigma} S_{g,n}$$

We make use of this expression to describe Guénault's [67G1] results on the effect of gold (type A impurity) and germanium (type B impurity) on S_g of silver. The ratio of the phonon–drag contributions for these two systems found experimentally is

$$\frac{S_g(A)}{S_g(B)} \sim 2$$

Theoretically,

$$\frac{S_g(A)}{S_g(B)} \sim \frac{(\sigma_n/\sigma)_A}{(\sigma_n/\sigma)_B}$$

and the experimental value of the right-hand side is ~ 1.7, in good agreement with the thermopower measurements and consistent with above arguments which would lead one to expect σ_n to decrease on adding type B impurity. Similar semiquantitative agreement is found for other alloy systems.

Summing up the situation, we find that at low temperatures S_g can be written

$$S_g = \sum_i \frac{\sigma_i}{\sigma} S_{gi}$$

where S_{gi} are determined by phonon-electron scattering and σ_i are determined by electron–impurity scattering. S_{gi} can be put in a form from which estimates of the contribution of the various regions can be made. The relative contributions of the positive and negative S_{gi} may be altered by changes in σ_i/σ produced by impurity scattering, and these changes are, in turn, a result of the character of the electron wave functions in the regions i and the character of the added impurity.

4.5 S_g at Intermediate Temperatures

At low temperatures the phonon–drag component typically increases in magnitude with temperature as the phonon density increases. At higher temperatures in pure metals phonon–phonon interactions begin to reduce the share of the phonon momentum passed on to the conduction electrons, and the phonon–drag thermopower decreases with temperature. At some temperature, usually ~ 0.1–$0.2\theta_D$, it is a maximum. For dilute alloys the phonons are also appreciably scattered by point defects; since Rayleigh-type scattering $\propto \omega^4$, its magnitude becomes greater as the temperature increases.

A study of S_g in the intermediate temperature region is made difficult because somehow the diffusion thermopower must be subtracted. However, generally, this is just the region of maximum confusion as far as the diffusion thermopower is concerned. In pure metals and more dilute alloys, electron–phonon scattering is appreciable compared with electron–impurity scattering. Consequently, the simple $S_d = aT$ relation is no longer satisfactory, because, on the one hand, electron-impurity scattering is no longer dominant and, at the same time, the temperature is considerably less than θ_D so that electron–phonon scattering cannot be considered elastic. That is, the temperature is considerably below that at which the linear dependence of thermopower on temperature due to the predominance of electron scattering by high-q phonons can be expected. Perhaps the best method of subtracting the diffusion thermopower is that suggested by Huebener [64H1]. The diffusion thermopower may be written as

$$S_d = \frac{W_i}{W_i + W_r} S_d^i + \frac{W_r}{W_i + W_r} S_d^r$$

The change in S_d on alloying, ΔS_d, is given by

$$\Delta S_d = S_d - S_d^i$$

$$= \frac{W_r}{W_i + W_r} (S_d^r - S_d^i)$$

$$= \frac{\rho_r / L_0 T}{\rho_i / LT + \rho_r / L_0 T} (S_d^r - S_d^i)$$

$$= \frac{\rho_r}{(L_0 / L)\rho_i + \rho_r} (S_d^r - S_d^i)$$

$$= \frac{\rho_r}{L_0 T / \kappa_i + \rho_r} (S_d^r - S_d^i) \tag{4.38}$$

where κ_i is the thermal conductivity of the pure metal. Thus

$$S_d = S_d^i + \rho_r \left(\rho_r + \frac{L_0 T}{\kappa_i}\right)^{-1} (S_d^r - S_d^i) \tag{4.39}$$

To apply this equation one is forced to invoke several assumptions. (a) At some high temperature, typically room temperature, it is assumed that the phonon–drag thermopower is negligibly small. This would be a poor approximation for aluminum [67G2], a moderately good one for copper [60B2], and a very good one for silver [65S1]. (b) S_d^i is then assumed to vary linearly with T as in free-electron theory. This is a rather bad assumption because generally in most metals there is no extended region in which $S_d \propto T$. Furthermore, for a pure metal it is only true theoretically for $T > \theta_D$ when phonon scattering is quasi-elastic. Despite this gloomy outlook there are a variety of results which indicate that linearity is a good approximation from room temperature down to ~ 150 K for silver and gold [65S1, 71G2, 73R1]. (c) S_d^r is also assumed to vary linearly with T, which again from the same references appears to be a good approximation for many noble metal alloys. It is also usually determined from a knowledge of S_d, S_d^i, ρ_r, and κ_i at room temperature.

Thereafter, $S_d^i(T)$, $S_d^r(T)$ (derived as outlined above) and $\kappa_i(T)$, ρ_r (measured experimentally) are substituted into Eq. (4.39) and $S_d(T)$ is calculated. The extrapolation is the most dubious when S_d^i ceases to be proportional to T. For alloys the equation will be better in this respect since elastic impurity scattering will play an increasing part as the impurity concentration increases. However, the theory above is based on S_d^r being independent of concentration, which certainly is not true for very concentrated alloys [64C1].

Generally speaking then, the precise separation of diffusion and phonon-drag thermopower is not easily accomplished, and for this reason the accuracy of an estimate of phonon-drag thermopower is generally limited to $\sim 0.05\ \mu$V/K at best. At high concentrations it will probably be worse. Having made the separation, one can proceed with the analysis only by further approximations. To exemplify these we refer the reader to Huebener's work on the change in phonon–drag thermopower caused by lattice vacancies in gold [64H1]. The main approximations are the use of the Debye theory for the lattice vibrations and free-electron theory for the electrons, and assuming a Raleigh-scattering term for the scattering of phonons by vacancies of the form

$$\tau_{pi}^{-1} = a\omega^4 \tag{4.40}$$

The parameter a is determined by fitting the theoretical results at 40 K. Perhaps the most interesting part of the work is the comparison of a with the theoretical expression for phonon scattering by point defects [55K1]. This calculation was carried out by Klemens for a cubic monatomic crystal with imperfections, neglecting dispersion of the phonons. The result is

$$a = (3\Delta_0 c / \pi v_c^3) L^2 \tag{4.41}$$

$$L^2 = \frac{1}{12} \left(\frac{\Delta M}{M} \right)^2 + \left[\frac{1}{\sqrt{6}} \frac{\Delta F}{F} - \left(\frac{2}{3} \right)^{\frac{1}{2}} Q\gamma \frac{\Delta R}{R} \right]^2 \tag{4.42}$$

where Δ_0 is the atomic volume of the crystal, c the mole fraction of the point defects, v_c the sound velocity, and γ the Grüneisen constant. M is the atomic mass of the crystal, F the force constant of a linkage, R the nearest neighbor distance, and ΔM, ΔF, and ΔR are the changes in these quantities at the location of the point defect. Q is a constant containing the contribution to the scattering matrix from the strains in the lattice outside the six nearest neighbors of the point defect. The Klemens theory yields a value for a 20 times smaller than Huebener's experimental result. The $\Delta M/M$ term alone gives a 40 times smaller. On the other hand, if Carruthers' [61C1] theory of strain-field scattering is used, then the strain field itself is sufficient to account for Huebener's result. The general conclusion is that anisobaric scattering, i.e., the $\Delta M/M$ term, is relatively unimportant in this case. Bhandari and Verma [67B6] have performed further theoretical work on this system.

4.6 S_g of Alloys

In the early history of phonon-drag thermopower it was commonly believed that the phonon-drag TEP of alloys would diminish rapidly as the solute concentration was increased. The early work of Blatt and Kropschot [60B2] concentrated on the $\Delta M/M$ term of Eq. (4.41). Studying a series of approximately 1 at.% alloys of Cu with neighboring solutes Zn and Ge, and with solutes neighboring Ag–, Cd, In, Sn, Sb–, they came to the conclusion that the phonon-drag thermopower of the 1% alloy $S_g^A \approx S_g^{Cu}$ for alloys Cu–Zn and Cu–Ge, whereas $S_g^A \approx 0$ for Cu–Cd, Cu–In, Cu–Sn, and Cu–Sb. These results are consistent with the anisobaric scattering being of primary importance. However, subsequent work by Lee [65L1] indicated that the magnitude of the phonon-drag TEP in alloys cannot always be described only in terms of scattering by atoms of different masses. For example, for dilute alloys of In in Sn [65F2], Pd in Ag, or Ni in Cu [65S1], phonon drag is suppressed very effectively despite the fact that $(\Delta M/M)^2$ is small. On the other hand, alloying Al and Si with Cu [65W3] produces a comparatively small suppression despite the fact that $(\Delta M/M)^2$ is large. Clearly, allowance must be made for the strain term and the force term in the Klemens equation. Schroeder and Lee [72L2] move a little in this direction in the following way. Let $S_g^0(T)$ be the phonon-drag thermopower of the pure metal at temperature T if the phonons transferred momentum to electrons only. Then the true phonon-drag TEPs of the pure metal and of the alloy are

$$S_g^P = S_g^0 \left[\frac{1}{\tau_{pe}} \middle/ \left(\frac{1}{\tau_{pe}} + \frac{1}{\tau_{pp}} \right) \right] \tag{4.43}$$

$$S_g^A = S_g^0 \left[\frac{1}{\tau_{pe}} \middle/ \left(\frac{1}{\tau_{pe}} + \frac{1}{\tau_{pp}} + \frac{1}{\tau_{pi}} \right) \right] \tag{4.44}$$

where τ_{pe}, τ_{pp}, and τ_{pi} are phonon relaxation times due to phonon–electron, phonon–phonon, and phonon–impurity scattering, respectively. Equations (4.43) and (4.44) are, of course, very crude approximations which neglect the many complicating features discussed in previous sections. Assuming that, as implied by these expressions, S_g^0 is not modified as a result of alloying, it then follows that

$$S_g^P/S_g^A = 1 + \left(\frac{1}{\tau_{pi}} \right) \middle/ \left(\frac{1}{\tau_{pe}} + \frac{1}{\tau_{pp}} \right) \tag{4.45}$$

Hence, a plot of S_g^P/S_g^A vs. $1/\tau_{pi}$ should be a straight line, as should a plot of S_g^P/S_g^A vs. $(\Delta M/M)^2$ if strain-field scattering is negligible. Schroeder and Lee's results on a series of 1% silver alloys are illustrated in Fig. 4.7. Despite the uncertainties in extracting the phonon-drag component of the TEP and the crudities of the above reasoning, it is quite clear that the solvents separate into two groups—those adjacent to silver and those adjacent to copper.

To incorporate the contribution of strain-field scattering to $1/\tau_{pi}$, the second term in (4.42) was replaced by $A(\Delta R/R)^2$ and ΔR estimated as equal to $100\Delta r/2c$, where Δr is the difference between the lattice parameter of an alloy of 1 at.% concentration and of pure silver, and c is the concentration in at.%. Figure 4.8 shows a plot of S_g^P/S_g^A vs. $1/\tau_{pi}$, including now the effect of lattice distortion empirically by treating A and a as adjustable parameters. At best these results indicate that anisobaric scattering does not necessarily predominate and that in some situations strain-field scattering is important. If we take the results at face value, we see from Table 4.6 that in this alloy system strain-field scattering predominates for the solutes adjacent to silver

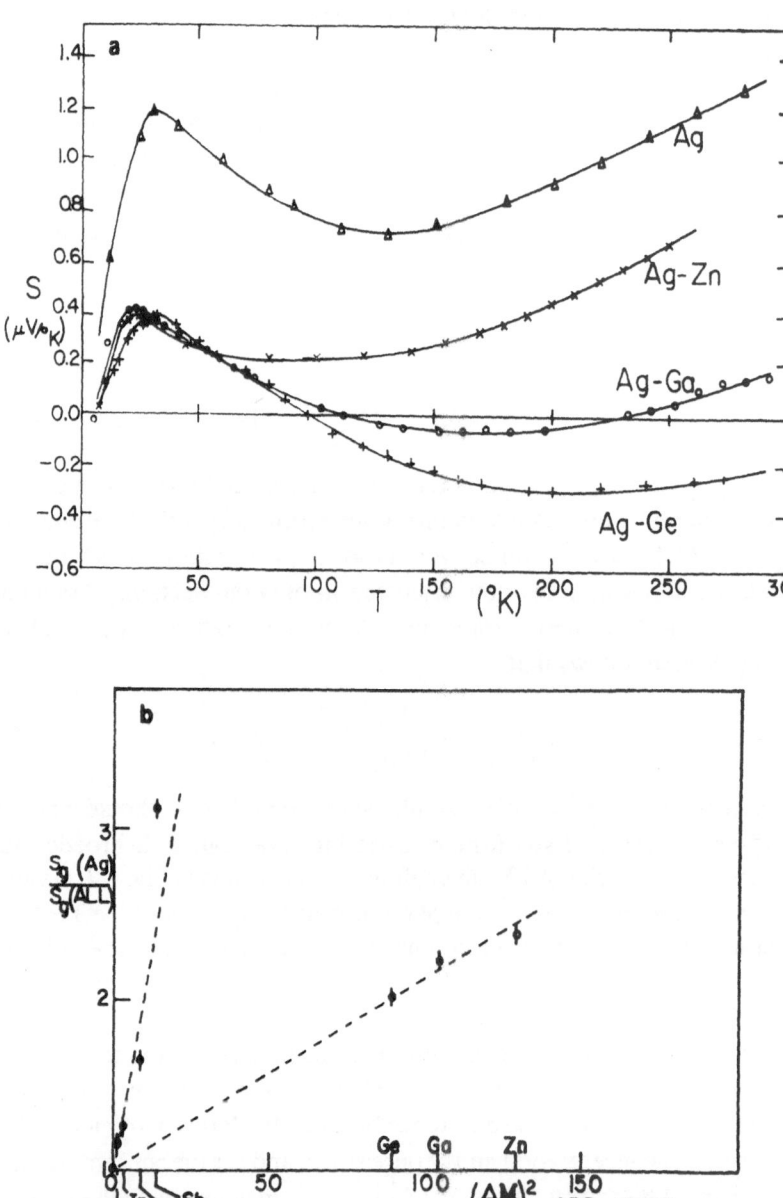

Fig. 4.7. (a) The thermoelectric power of 1 at.% alloys of silver with solutes adjacent to copper. (b) Ratio of the peak values of S_g for pure silver and silver alloys (from [72L2]).

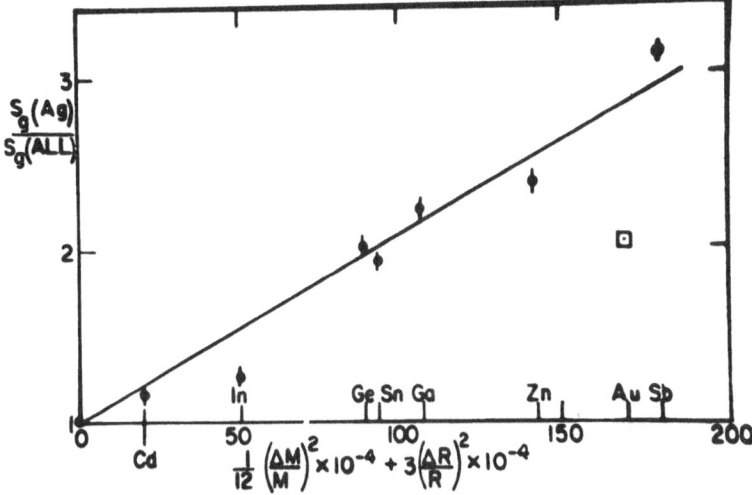

Fig. 4.8. Ratio of peak phonon-drag thermopowers for pure silver and for silver alloys plotted as a function of phonon-scattering parameters (from [72L2]).

and anisobaric scattering predominates for solutes adjacent to copper in the periodic system. Even so there are still exceptions; for example, Ag–Pd has a phonon-drag peak that is much smaller than expected [65S1].

There is the further problem that in many alloy systems the phonon-drag thermopower persists to high concentrations. As an example, we show in Fig. 4.9a the results for the Cu–Zn and Cu–Ga systems [64C1], in which phonon drag not only persists up to the α-phase boundary but even shows an increase in magnitude at the higher concentrations. Similar results hold for

Table 4.6. Mass and Strain Factors in Dilute Alloys

Alloy	$1/12(\Delta M/M)^2 \times 10^4$	$3(\Delta R/R)^2 \times 10^4$
Ag–Cd	1.5	18.9
Ag–In	3.5	47.4
Ag–Sn	8.4	85.8
Ag–Sb	13.8	165.6
Ag–Zn	129.4	12.9
Ag–Ga	104.3	2.7
Ag–Ge	89.2	0.3
Ag–Pd	0.2	24.6
Ag–Au	170.0	0.073

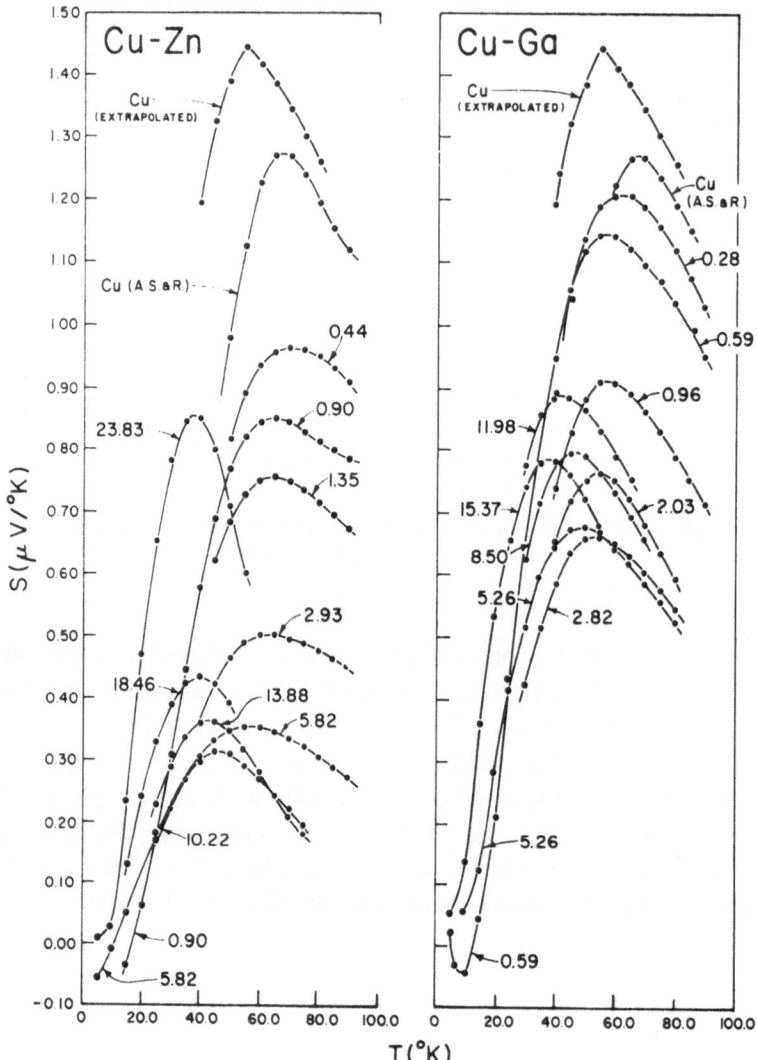

Fig. 4.9a. The thermopowers in the phonon-drag region of the solid solutions of zinc and gallium in copper (from [64C1]).

the Ag–Cd system [73A2, 73R1]. For alloys of Ga, Ge, and As in Cu [64C1], and Zn in Ag [74H1], phonon drag persists up to the α-phase boundary (Fig. 4.9b). For all these alloys $\Delta M/M$ is relatively small and perhaps the persistence of the phonon-drag peak is understandable. However, it is surprising to find that for Ag–Au phonon drag is apparently present over the

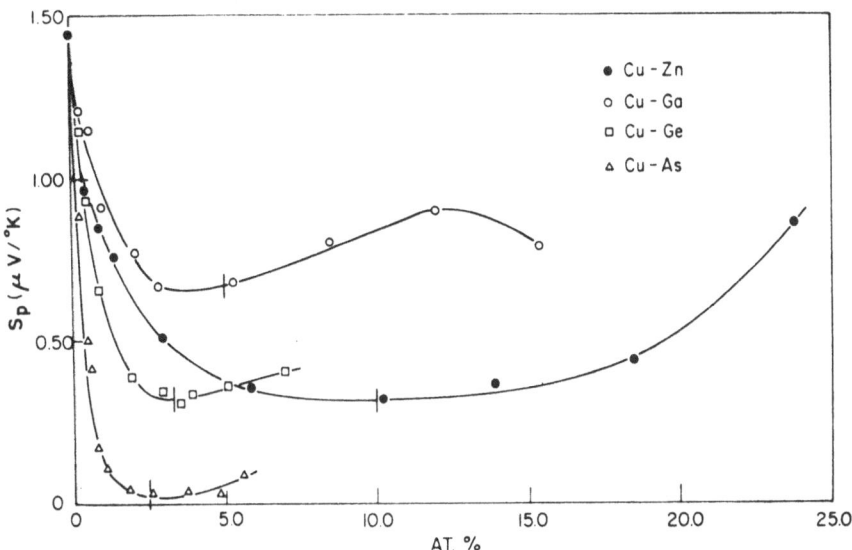

Fig. 4.9b. The thermopowers of the phonon–drag peak heights of the solid solutions of zinc, gallium, germanium, and arsenic in copper (from [64C1]).

whole concentration range from 0 to 100 at.% Au [70C2] as shown in Fig. 4.10.

The presence of impurities diminishes not only S_g but also κ_g, the lattice thermal conductivity. Since κ_g is proportional to τ_p [see Eq. (2.34)], it follows that

$$(S_g^P/S_g^A) = (\kappa_g^P/\kappa_g^A) \tag{4.46}$$

The measured changes of κ_g and S_g due to alloying [64B2] do not agree very well with the simple relationship (4.46). This failure is not particularly surprising in view of the following considerations.

First, even assuming the most ideal situation (spherical Fermi surface and a dispersionless (Debye) spectrum for the phonons), high-frequency, large-$\mathbf{q}$ phonons are weighted differently in the integrations over the phonon spectrum which give the expressions for S_g and κ_g. Consequently, collision processes, such as Rayleigh scattering, which have a strong wavelength dependence, will reduce S_g and κ_g by different relative amounts. In other words, in the averaging process over the phonon distribution $\langle \tau_{pi} \rangle_{S_g} \neq \langle \tau_{pi} \rangle_{\kappa_g}$.

Second, whereas in the expression for the thermal conductivity the product $\hbar \omega \mathbf{v}_g$ appears, where $\mathbf{v}_g$ is the phonon group velocity, the phonon-drag effect involves the exchange of crystal momentum $\hbar \mathbf{q}$. Consequently,

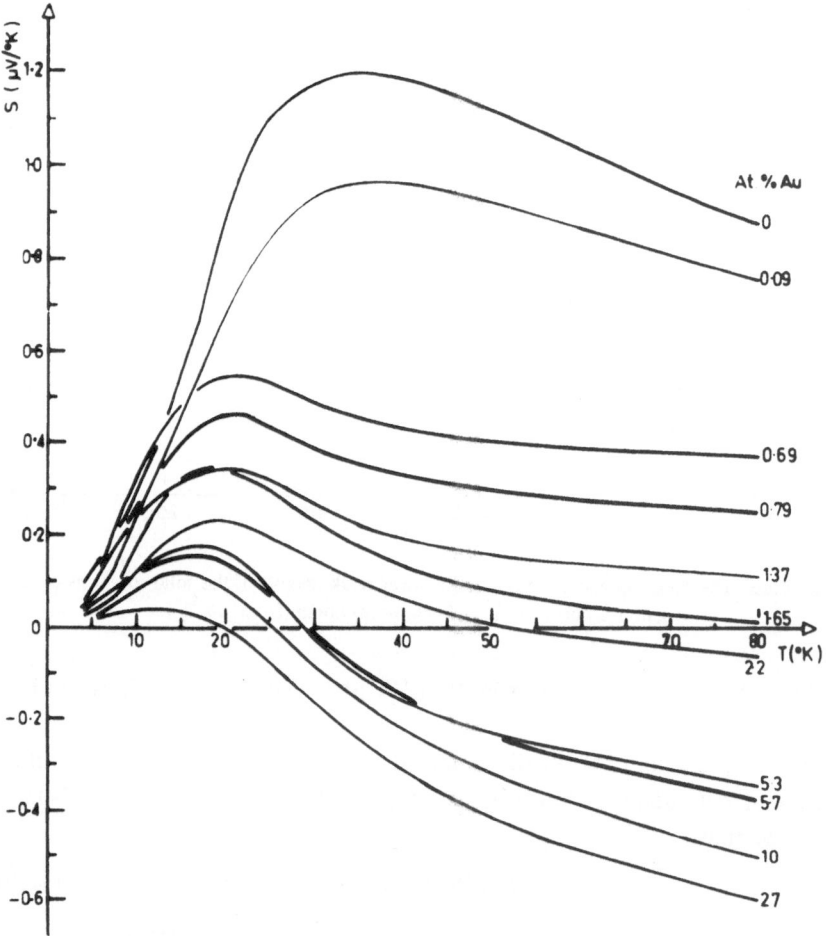

Fig. 4.10a. The thermopower of the silver–gold alloy system between 4 and 80 K (from 70C2]).

dispersion of the phonon spectrum will have a more pronounced effect on κ_g than on S_g, because regions of the phonon spectrum where v_g is vanishingly small (ω vs. $\mathbf{q}$ curve nearly flat) do not contribute to κ_g but do contribute to S_g. These regions in the phonon spectrum generally occur for relatively large $\mathbf{q}$, where Rayleigh scattering is most effective.

In passing we note that for the α-phase Cu–Zn and Ag–Cd alloys the thermopower, in general, and the phonon-drag thermopower, in particular, are for some concentrations strongly dependent on the thermal history of

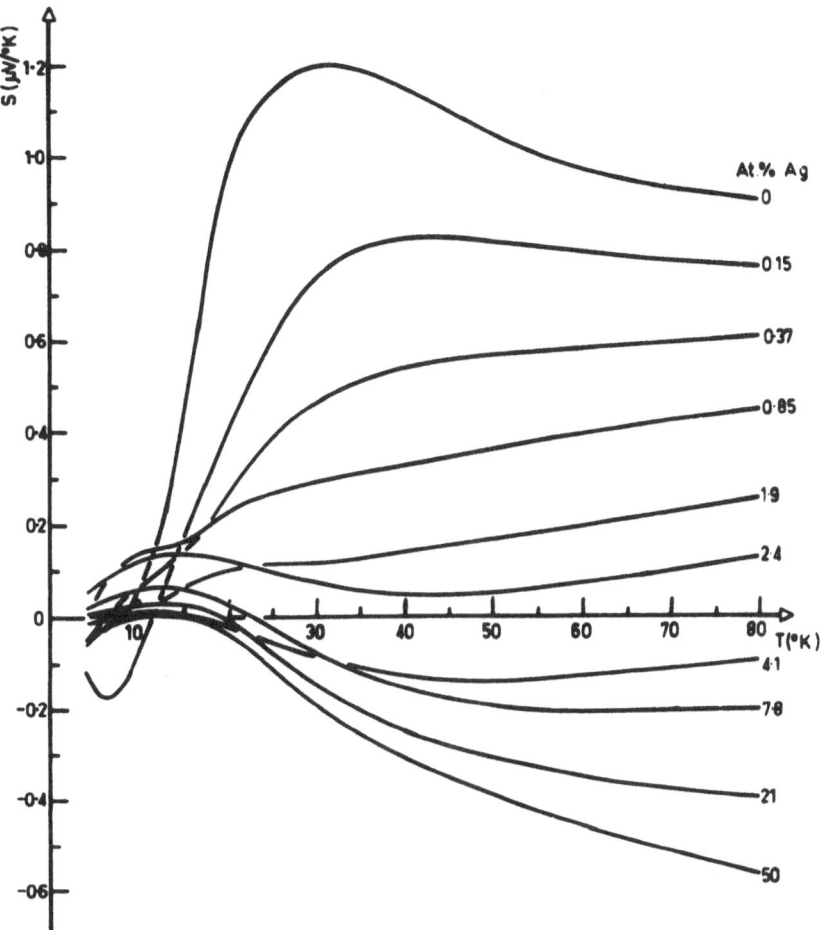

Fig. 4.10b. The thermopower of the gold–silver alloy system between 4 and 80 K (from [70C2]).

the specimen. We illustrate this in Fig. 4.11 for Ag–Cd. The work of Waldman and Bever [72W1] on the effect of quenching on the electrical resistivity shows that samples of Ag–Cd quenched from 600°C are more disordered than those slowly cooled. The rather peculiar conclusion is that disorder in this material leads to a larger phonon-drag thermopower. Electrical resistivity measurements have not been performed on the Cu–Zn alloys, but the thermopower results are similar and even more pronounced than those in Ag–Cd. If one ascribes the changes in phonon-drag TEP to

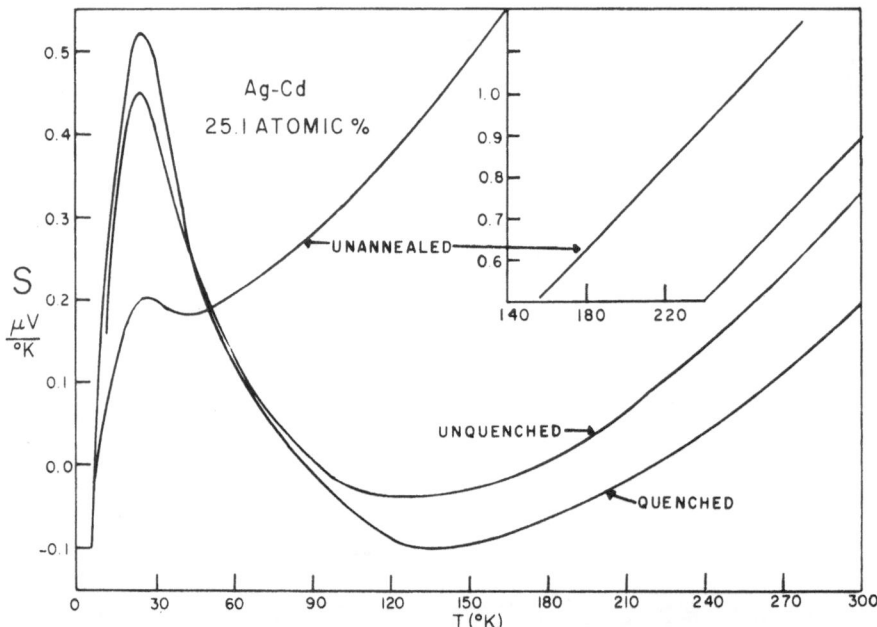

Fig. 4.11. Effect of annealing and quenching on the thermopower of a Ag + 25.1 at.% Cd alloy (from [72W1]).

phonon-scattering effects, then one would conclude that the negative component of phonon drag has decreased more than the positive to give an apparent increase in the positive phonon-drag peak.

4.7   Phonon Drag or Phony Phonon Drag?

We shall conclude this chapter with a discussion of another contribution to the thermopower which for pure metals theoretically has the same low-temperature T^3 dependence and high-temperature T^{-1} dependence as the simple phonon-drag theory but does not require the presence of a phonon flux. Furthermore, the predicted magnitudes are similar to those observed experimentally. Here we shall consider the predictions, what evidence there is for the effect, and whether in the past phonon-drag effects have been confused by its presence.

The theory has been developed by P. E. Nielsen and P. L. Taylor [68N1, 70N1, 74N1] for pure metals and dilute alloys. They considered a simple model in which free electrons were scattered by phonons or impurities and calculated the second-order corrections to the scattering probabilities for these processes. According to Migdal's theorem [58M4], such renormalizations are very small, changing the scattering probabilities by a factor of the order of $1+m/M$, where m/M is the ratio of electron mass to ion mass. However, as pointed out by Nielsen and Taylor, it is the energy dependence of the scattering probability in the vicinity of the Fermi energy which is important for the determination of thermopower. Thus, second-order corrections which are strongly energy dependent near the Fermi level can contribute significantly to the thermopower.

4.7a *Pure Metals*

Let us first consider the Nielsen–Taylor theory for pure metals in which only electron–phonon scattering processes are involved. Processes which add coherently, in that they connect the same initial and final state, are illustrated by the diagrams shown in Fig. 4.12. Figure 4.12a represents the first-order terms usually considered in the theory of electron–phonon scattering in which a phonon is absorbed or emitted. The diagrams in (b) and (c) are symbolic of a number of second-order processes in which the intermediate electron and phonon states can either be particles or holes. Nielsen and Taylor show these second-order terms to be highly energy dependent and therefore significant in calculating the thermopower.

The results of their calculations can be expressed in terms of the thermopower parameter x, which is defined by

$$S = S_0 x$$

where

$$S_0 = \frac{\pi^2 k_B^2 T}{3e\eta} \tag{4.47}$$

In their calculations they neglect Umklapp processes and obtain for the thermopower parameter

$$x = x_1 + \Delta x_1 \tag{4.48}$$

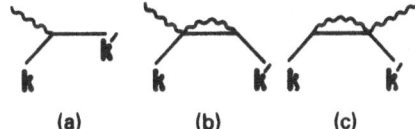

Fig. 4.12. Second-order processes coherent with first-order electron–phonon scattering.

(a)   (b)   (c)

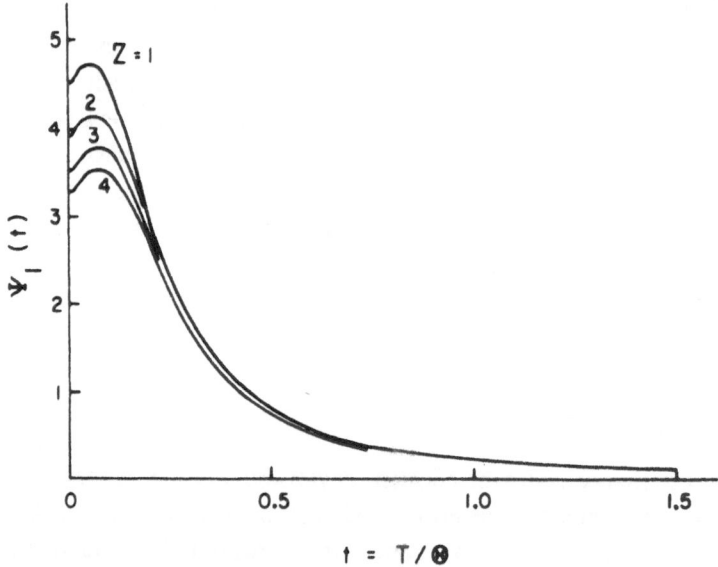

Fig. 4.13. Temperature dependence of the contribution to the thermoelectric parameter x of virtual phonons in a pure metal for various values of the valence Z. The temperature is measured in units of the Debye temperature θ_D (from [74N1]).

where x_1 is the well-known temperature independent first-order contribution

$$x_1 = 3 - 2 \left(\frac{\partial \ln |V|}{\partial \ln \varepsilon} \right)_{\varepsilon = \eta}$$

and the second-order contribution Δx_1 is given by

$$\Delta \dot{x}_1 = \frac{\eta N V}{k^2 \theta_D^2} \frac{m}{M} \psi_1 \left(\frac{T}{\theta_D} \right)$$

In these expressions, V is the $q = \frac{2}{3} k_f$ component of the Fourier-transformed screened ionic pseudopotential, N is the number of ions, η is the Fermi energy, k is Boltzmann's constant, θ_D is the Debye temperature, and $\psi_1(T/\theta_D)$ is a complicated function of temperature shown in Fig. 4.13. Since the potential V is negative and S_0 is negative, the term Δx_1 results in a positive contribution to S. The behavior of $S_0 \Delta x_1$ is similar to that associated with the phonon-drag effects discussed at the beginning of this chapter. At low temperatures $S_0 \Delta x_1$ departs from linearity as $T^3 \ln T$, at high tempera-

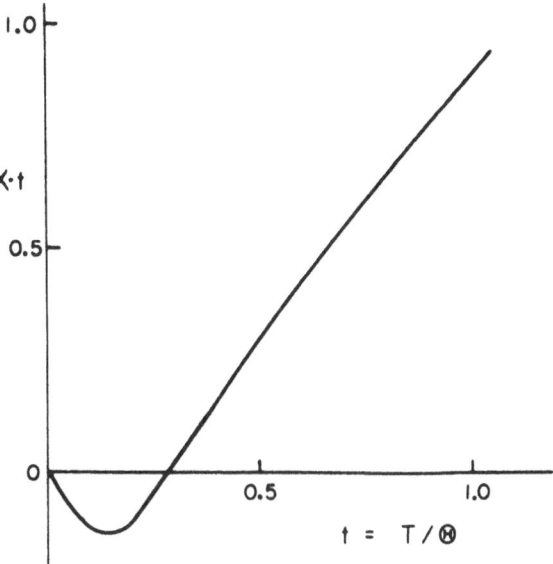

Fig. 4.14. Typical form of the negative of the Seebeck coefficient S to be expected in a pure material in the absence of phonon drag. Here $x(T/\theta_D)$, which is proportional to $-S$, is plotted for the case where $x = 1 - \frac{1}{2}\psi_1(T/\theta_D)$ in a monovalent metal (from [74N1]).

tures behaves as $1/T$, and has a maximum between $0.1\theta_D$ and $0.2\theta_D$. Because of this temperature dependence, the Nielsen–Taylor effect for pure metals is often referred to as "phony phonon drag." A plot of $x \cdot (T/\theta_D)$ vs. (T/θ_D) for a monovalent metal is shown in Fig. 4.14.

4.7b Dilute Alloys

For dilute substitutional alloys there are a number of scattering terms which contribute to the thermopower. In addition to the scattering of electrons by phonons, there will be elastic and inelastic scattering by impurities. In general, since inelastic scattering processes conserve neither energy nor momentum, their amplitudes will not be coherent with either electron–phonon scattering or elastic impurity-scattering amplitudes. For this reason, the contribution of the inelastic scattering processes to the thermopower will be negligible.

The first- and second-order elastic scattering processes calculated by Nielsen and Taylor are represented in Fig. 4.15. The first-order process is

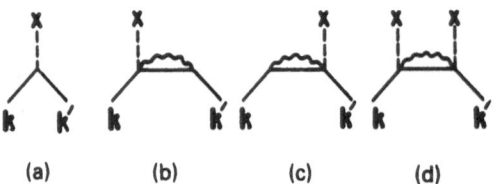

Fig. 4.15. First- and second-order elastic impurity scattering events (from [74N1]).

represented by (a), and (b) through (d) represent the possible second-order processes coherent with (a). From their calculations, Nielsen and Taylor obtain three additional terms which contribute to the thermopower of dilute alloys. There is the usual first-order scattering contribution

$$\Delta x_2 = 1 - 2\left(\frac{\partial \ln |U|}{\partial \ln \varepsilon}\right)_{\varepsilon = \eta}$$

and two additional second-order contributions

$$\Delta x_B = 6\frac{\eta NU}{(k\theta_D)^2}\frac{m}{M}\frac{n_0}{N}\psi_2\left(\frac{T}{\theta_D}\right)$$

$$\Delta x_C = \frac{\eta NV}{(k\theta_D)^2}\frac{m}{M}\left(\frac{2N}{n_0}\right)^{1/3}\psi_3\left(\frac{T}{\theta_D}\right)$$

In these expressions, U is the $\mathbf{q} = \frac{2}{3}k_f$ component of the Fourier-transformed impurity-scattering potential, n_0/N is the valence, and the functions ψ_2 and ψ_3 are shown in Figs. 4.16 and 4.17, respectively. At low temperatures ψ_1 and ψ_2 both approach unity. Note that while Δx_C is inherently negative, Δx_B will be positive or negative depending on whether U is positive or negative.

The lattice and impurity contributions to the thermopower can now be combined as follows to give the total thermopower. Since the contributions of independent scattering processes to the total thermopower are approximately proportional to their respective contributions to the total electrical resistivity, the total thermopower parameter can be written

$$x = \frac{\rho_i(x_1 + \Delta x_1) + \rho_r(x_2 + \Delta x_B + \Delta x_C)}{\rho_i + \rho_r} \tag{4.49}$$

where ρ_i and ρ_r are, respectively, the lattice and impurity contributions to the resistivity. The parameters x_1 *and* Δx_1 are the same as those defined in Eq. (4.48). At the lowest temperatures the impurity term will predominate

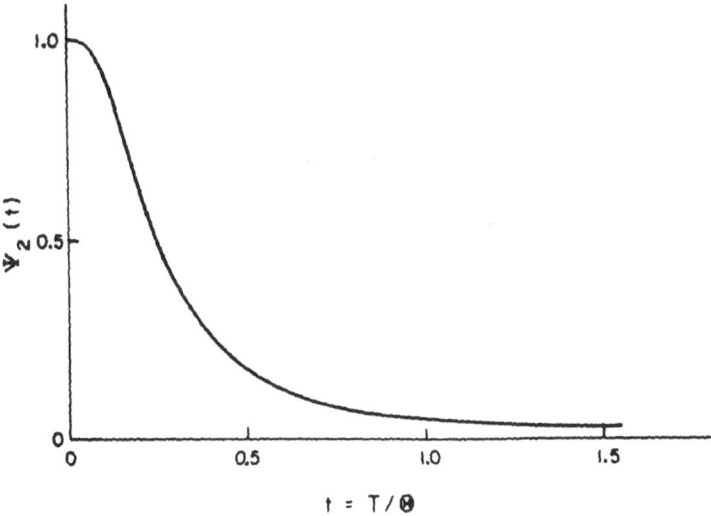

Fig. 4.16. Temperature dependence of the contribution to the thermoelectric parameter x of processes in which an electron is twice scattered by an impurity ion exhibiting virtual recoil (from [74N1]).

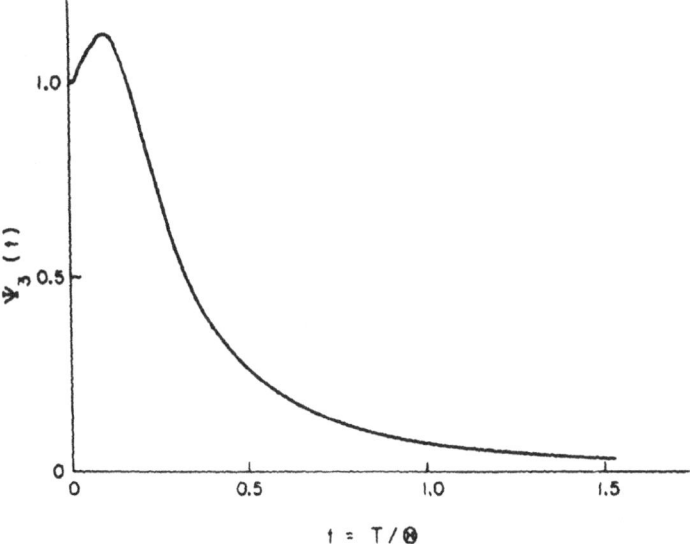

Fig. 4.17. Temperature dependence of the contribution to the thermoelectric parameter x of processes in which an electron is scattered both by an impurity exhibiting virtual recoil and in a wave-number-conserving phonon interaction (from [74N1]).

to give $S \propto T$ with the constant of proportionality

$$\frac{\pi^2 k^2}{3e\eta} [x_2 + \Delta x_B(0) + \Delta x_C(0)] \tag{4.50}$$

and, as we saw, this will be influenced quite markedly by the sign of U. At high temperatures, on the other hand, $S \propto T$ with the constant of proportionality

$$\frac{\pi^2 k^2}{3e\eta} (x_1) \tag{4.51}$$

which is quite different from the low-temperature slope. In between, peaking at $T \sim 0.15 \theta_D$, is the phony phonon-drag term $S_0 \Delta x_1$.

4.7c   Evidence for "Phony Phonon Drag"

It is evident that one of the most interesting regions for experimental investigation is at low temperatures $\gtrsim 4.2$ K, where the linear term can be investigated in detail. We shall now consider what effect the solute has on the low-temperature slope. It is primarily U in x_2 and Δx_B which is of importance here. This will depend markedly on the valence difference between solvent and solute atoms. Nielsen and Taylor show that the addition of the terms Δx_B and Δx_C makes a substantial improvement in understanding the results of Guénault and MacDonald [61G1] on dilute alloys of Na, Rb, and Cs in K. They also consider in detail [70N2] the measurements of the thermopower of Pd–H alloys by R. Fletcher, N. S. Ho, and F. D. Manchester [70F3] which cannot be explained in terms of impurity-induced s–d scattering.

Let us next consider the phony phonon-drag term Δx. For the pure monovalent metals Nielsen and Taylor estimate contributions to the thermopower from this term to range from 0.19 μV/deg^{-1} in silver to 1.74 V/deg^{-1} in lithium. This is sufficient to make a very appreciable difference in the results previously ascribed to phonon drag. Is there a way of separating the two or showing conclusively that the two coexist in the pure metal? One way would be the presence in a metal of two extrema. According to Nielsen and Taylor the maximum in the phony phonon-drag peak should be positive for a free-electron metal and should occur at $\sim 0.15 \theta_D$ and, coincidentally, peaks in the measured thermopower should be in the range 0.1 to 0.2 θ_D. As yet no resolvable double peaks have been observed, but many pure metals show double extrema of opposite sign. For example, polycrystalline tin in which $\theta_D = 150$ K has a negative TEP below 12 K and a positive peak at

30 K [65F2]. The group II metals, Mg, Zn, and Cd [70R1], show both negative dips followed by positive peaks and positive peaks followed by negative dips. These occur in the regions below 0.2 θ_D and are ascribed by the authors to contributions to the phonon-drag thermopower by various sheets of the Fermi surface. In aluminum, where the low-temperature T^3 and high-temperature T^{-1} law holds very well [67G2], the "phony phonon-drag" contribution is negative. However, negative phony phonon-drag thermopowers may well be possible when the details of the Fermi surface are taken into account. In lead, which has a low θ_D and which might therefore be expected to give a substantial phony phonon drag, the thermopower peak is weak and negative. Altogether there appear to be no experimental data in which a distinctive phony phonon-drag contribution is clearly evident. The nearest are the alkali metals as discussed by Nielsen and Taylor.

When we come to consider alloys, according to Eq. (4.49), the phony phonon-drag contribution is reduced by a factor $\sim\rho_i/(\rho_i+\rho_r)$. On the other hand, we expect the pure metal phonon-drag contribution to be reduced by the factor $\tau_{pi}/(\tau_{pe}+\tau_{pi})$ [Eq. (4.44)—neglecting phonon–phonon interactions for the moment]. The τ_p are phonon-relaxation times. This leads to the possibility of making alloys such that the effect of reducing τ_{pi} (impurity scattering) is considerably more than the effect of increasing ρ_r and vice versa, and possibly reducing one of the components until the other becomes clearly visible. We have seen that one way of reducing τ_{pi} is by adding impurities of substantially different mass from the solvent atoms. On the other hand, impurity scattering of electrons is roughly proportional to the square of the valence difference between solute and solvent atoms.

Hence, perhaps the obvious system to look at is the Ag–Au system. We have already remarked that the "phonon-drag" peak in this system initially decreases rapidly with concentration and, thereafter, remains relatively independent of concentration (see Fig. 4.18). Is this possibly caused by the real phonon drag in Ag being rapidly quenched by scattering of phonons by the Au, leaving the phony phonon-drag contribution only at higher concentrations? The answer is probably not, and for the following reasons. The Nielsen–Taylor "phony phonon-drag" term is actually a component of the diffusion thermopower associated with phonon scattering. Denoting, as before, characteristic lattice thermopower by S_d^i and the characteristic thermopower associated with impurity scattering by S_d^r, the measured thermopower is

$$S = \frac{W_r S_d^r + W_i S_d^i}{W_r + W_i}$$

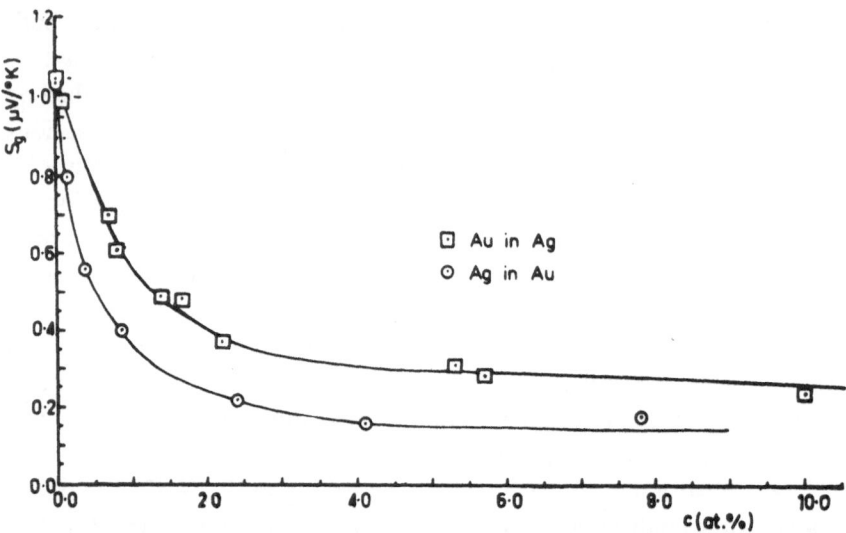

Fig. 4.18. Value of the phonon-drag peak in the Ag–Au alloy system (from [70C2]).

Estimates of W_r and W_i can be made from the results of Crisp and Rungis [70C2]. For temperatures near that of the thermopower peak we find

C (at.%)	W_r/W_i, $T=30$ K	W_r/W_i, $T=50$ K
0.09	~ 1	0.25
0.69	~ 3	1.2
2.2	~13	4
10	~47	11.5

At 30 K it is clear that the diffusion thermopower associated with lattice scattering will decrease very appreciably as the concentration is increased. If the phony phonon-drag mechanism is to be observed, it would appear that it will be for moderate temperatures $\sim \theta_D/5$ and for low-concentration alloys—say less than 2 at.%. For the Ag–Au system, the thermopower does indeed decrease rapidly for concentrations up to 2 at.% but not in such a way as to be dramatically different from that of other alloy systems. Guénault [71G2] originally used the above argument in studying a number of measurements on pure and dilute alloys at $T < 10$ K and came to similar conclusions.

Within the above range where it appears that the phony phonon-drag effect might be appreciable, Bourassa and collaborators [72D1, 73B1, 73R2] have performed calculations of the change in thermopower antici-

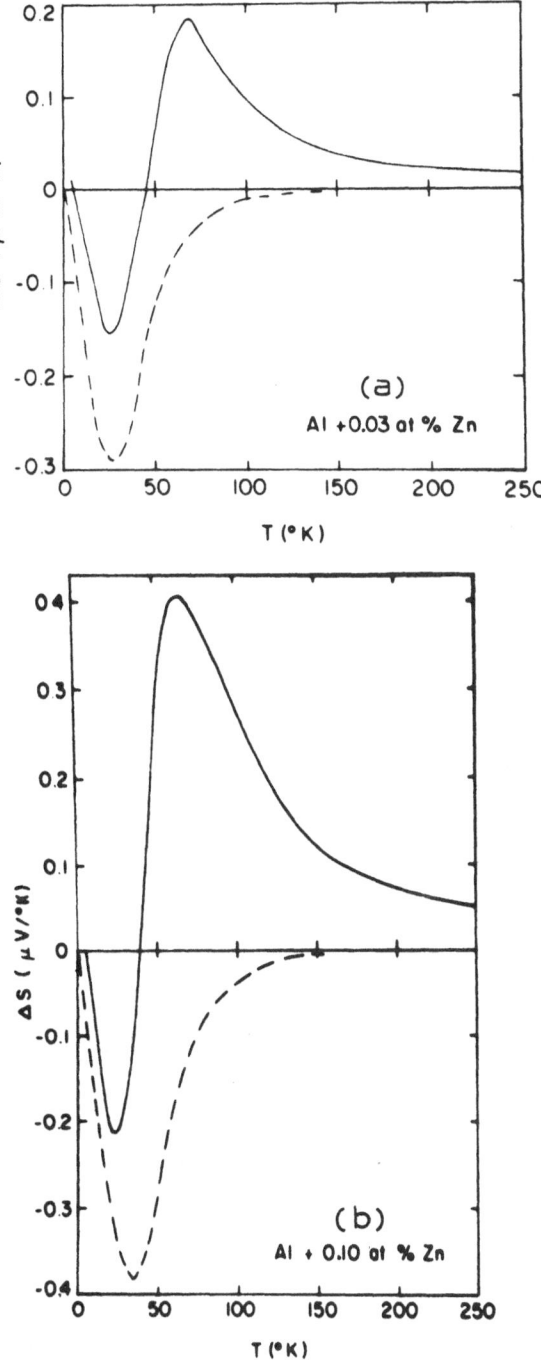

Fig. 4.19. The change in the thermopower of dilute Al–Zn alloys with respect to that of pure aluminum. Experimental results of Huebener are the solid curves and calculations using the Nielsen–Taylor theory are shown by the dashed curves. (a) Al+0.03 at.% Zn; (b) Al+0.1 at.% Zn (from [72D1]).

pated on alloying according to the Nielsen–Taylor theory. An illustration of their results for dilute Al–Zn alloys is shown in Fig. 4.19 (dashed lines) along with the change in thermopower found experimentally by Huebener [68H2]. Huebener's theory of these results is that the positive peak in ΔS is caused by a decrease in the magnitude of the negative phonon-drag peak in Al. The negative dip which corresponds to an increase in magnitude of the thermopower Huebener associates with anisotropic impurity scattering of the type suggested by Dugdale and Bailyn [67D1] for the noble metals. Similar results are obtained for Ag and Mg in Al. Dudenhoeffer and Bourassa agree with Huebener that the positive peak in the experimental results for ΔS probably corresponds with the conventional phonon-drag thermopower, but they suggest that the Nielsen–Taylor theory is an alternative explanation of the negative dip. Similarly, they have performed calculations for the Pb alloys with Tl, Bi, Cd, In, and Sn on which thermopower measurements were performed by Jan *et al.* [58J1] and Gold and Pearson [61G3], who concluded that they were consistent with conventional phonon drag only if the scattering of phonons by impurities depended critically on the valence difference between solvent and solute. A dependence on valence difference is indeed incorporated within the Nielsen–Taylor theory, and again Bourassa and Dudenhoeffer [73B1] obtain fits to the experiment which are good considering the approximations made and the difficulties in separating the various terms.

The overall conclusion is that the existence of the Nielsen–Taylor effect has been neither proved nor disproved. Perhaps we have to wait until more sophisticated calculations of the phonon-drag and the phony phonon-drag thermopowers can be performed. So far the phony phonon-drag calculations have been based on free-electron theory and have neglected Umklapp processes. Past experience with thermopower calculations suggests that improvement of these aspects of the theory could completely alter the calculated values. Certainly, however, the Nielsen–Taylor theory must be kept in mind in any interpretation of phonon-drag thermopower.

4.7d *Effects of Higher-Order Scattering Processes*

Since the work of Nielsen and Taylor on the effects of second-order scattering processes, Hasegawa [74H2, 74H3] has shown that third-order corrections to the electron-scattering probabilities should also play a significant role in determining thermopower. Examples of such third-order processes are displayed in Fig. 4.20 and can be compared with second-order

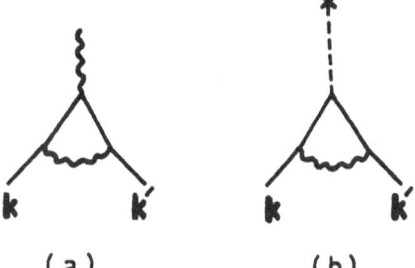

Fig. 4.20. Examples of third-order scattering processes.

(a)   (b)

processes shown in Figs. 4.12 and 4.15. For example, Fig. 4.20a contains a process in which an electron first emits the intermediate-state virtual phonon, then emits the final-state phonon, and finally reabsorbs the virtual phonon. This process is similar to a second-order process contained in Fig. 4.12b with the exception that the final-state phonon in the second-order process is emitted simultaneously with the intermediate-state virtual phonon.

In his calculations for both pure metals and dilute alloys, Hasegawa finds third-order terms in the scattering probability proportional to the Fermi function $f_0(\varepsilon)$, which are therefore strongly energy dependent near the Fermi level. His results give contributions to the thermopower comparable both in magnitude and temperature dependence to those of Nielsen and Taylor.

The total thermopower parameter x as given by Eq. (4.49), when modified to include the results of Hasegawa's calculations, can be expressed as

$$x = \frac{\rho_i(x_1 + \Delta x_1 + \Delta x_1') + \rho_r(x_2 + \Delta x_B + \Delta x_C + \Delta x_C)}{\rho_i' + \rho_r} \qquad (4.52)$$

where $\Delta x_1'$ results from third-order corrections to the electron–phonon interaction (Fig. 4.20a), and $\Delta x_C'$ results from third-order corrections to electron–impurity scattering (Fig. 4.20b). All of the remaining terms are the same as those defined in Eq. (4.49). The terms $\Delta x_1'$ and $\Delta x_C'$ are inherently negative and therefore add to the corresponding inherently negative Nielsen–Taylor terms, Δx_1 and Δx_C.

Hasegawa also examined the possibility of significant contributions coming from terms higher than third order in the scattering and found none. This can be understood, at least qualitatively, in the following way: For each interaction involving n phonons, a factor $(m/M)^{n/2}$ appears in the scattering amplitude (m denotes the electron mass and M the ion mass) and the corresponding terms in the scattering probability are therefore of order $(m/M)^n$. However, the logarithmic energy derivative of a product of terms containing strongly energy-dependent Fermi functions will only give rise to a sum of terms, each of order $(\eta/k_B\theta_D)^2$ (which is of order M/m). Therefore, for any scattering processes involving more than $n = 2$ phonons, $(m/M)^n(\eta/k_B\theta_D)2 \sim (m/M)^{n-1}$ will be negligibly small. At present, it appears that any higher-order scattering processes will not result in any new significant contributions to the diffusion thermopower.

5 | The Thermoelectric Power of Transition Metals

At any temperature the thermoelectric power of the 24 transition metals* is roughly an order of magnitude greater than that of "simple" metals. For this reason at least one limb of any thermocouple is generally a transition metal or alloy. It is also for this reason that the bulk of numerous measurements made on transition metals has been made with a view to determining their suitability and reliability as thermoelements, and not as a means of studying their electronic properties. Apart from the early work of Mott [36M1], who gave a measure of understanding as to why elements such as palladium and platinum had a large thermopower, the only scientific work of note has been carried out in the last ten years. Nevertheless, as we have said, for some transition metals there is no shortage of measurements.

5.1 Special Problems in Transition Metals

5.1a s- and d-Conduction

Transition metal atoms are characterized by vacant electronic states of both s- and d-character. Consequently, as solids they have overlapping

---

* Lanthanum, the heaviest group IIIB element, can also be classified as the first of the rare earths. However, it does not contain any electrons originating from atomic f-states and can be properly regarded as a transition metal.

s- and d-bands at the Fermi surface, and these vacant d-states are responsible for the many interesting properties of such metals.

Because the d-electrons lie rather closer to the atomic nuclei than the s-electrons, the d-bands in the solid are relatively narrow in energy. Furthermore, as they can accommodate five times as many electrons as an s-band, the density of states $N_d(\varepsilon)$ can be quite high, typically an order of magnitude greater than in a free-electron-like s-band. Yet another important feature of d-bands is that they are built of five sub-d-bands, formed from wave functions of different symmetries, so that a typical d-band contains a lot of "structure," and the density of states is not high at *every* energy in the band. As an example, we show in Figure 5.1 the calculated band structure of palladium.

As a consequence of this pronounced structure two points can be made. (a) The value of the density of states at the Fermi level, $N_d(\eta)$, may vary

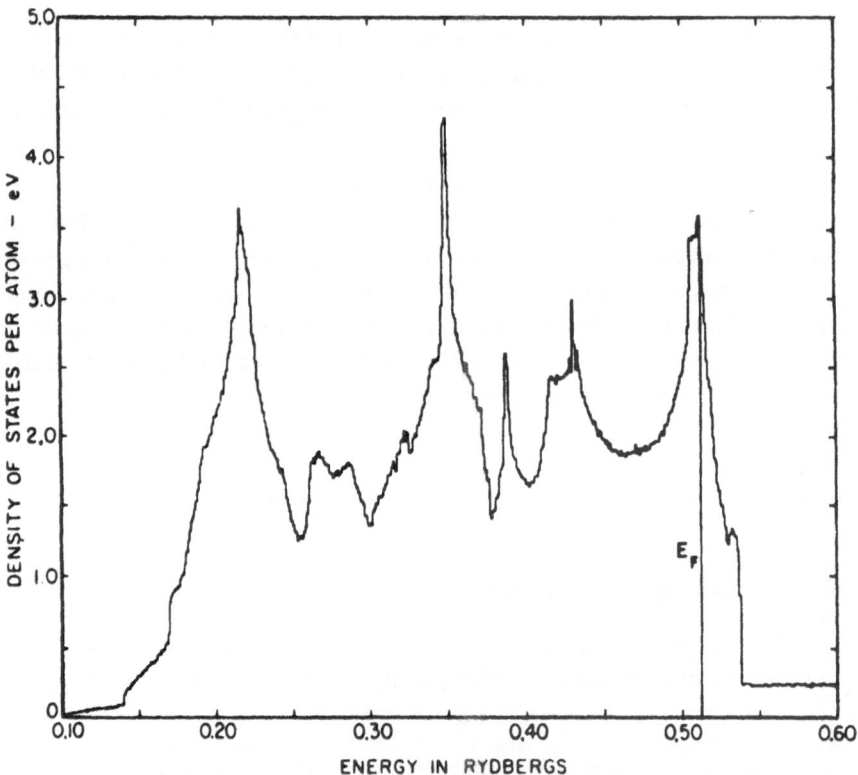

Fig. 5.1. The density of states in palladium as calculated by Mueller *et al.* (from [70M1]).

greatly from one transition metal to another, and this is reflected in the wide range of values for the observed specific heats of those metals [66G1]. (b) The energy *derivative* of the density of states at the Fermi level, $(dN_d(\varepsilon)/d\varepsilon)_{\varepsilon=\eta}$, may also be much higher than in simple metals. This latter point is of major importance in the study of diffusion thermopower.

As far as transport properties are concerned, the simplest model is to assume that the electrons can be separated into two groups; those originating from s-like atomic levels which, in the solid state, form a free-electron band, together with those in the d-band mentioned in the preceding paragraph. To this approximation the total electrical conductivity σ is therefore given by the sum of an s-component and a d-component, σ_s and σ_d, respectively, so that

$$\sigma = \sigma_s + \sigma_d \tag{5.1}$$

As a working hypothesis it is generally assumed that as a consequence of the high value of $N_d(\eta)$, $\sigma_d \ll \sigma_s$ and $\sigma \sim \sigma_s$. However, a recent detailed interpretation of the Hall coefficient of platinum [73G1], based on Fermi surface geometry that had been obtained by de Haas–van Alphen techniques, showed that the contribution to the conductivity from parts of the Fermi surface that are generally regarded as d-like could be comparable to σ_s. The simplifying assumption $\sigma_d \ll \sigma_s$ must therefore be viewed with care, although it must be emphasized that without such an approximation any progress would be very much more difficult.

5.1b *Electron–Electron Collisions*

In spite of the very high density of conduction electrons in metals, the contribution to the resistivity of "simple" metals from electron–electron collisions is negligible. There are two reasons for this: (a) When all the conduction electrons are alike, momentum is conserved during any electron–electron N-process so that such collisions do not in any way attenuate the total electric current. (b) On the other hand, it is possible for electrons to lose momentum to the lattice via electron–electron U-processes. However, the net effect of these processes is also very small, partly because the scattering cross section for screened Coulomb interactions, Σ_{SC}, is intrinsically small, and partly because the application of the exclusion principle requires that only electrons close to the Fermi energy take part in this process. In fact, in any scattering event both the initial and final electron states are confined to a narrow range of energy within kT of the Fermi energy, so that the effective electron–electron scattering cross section Σ_{ee} is given by

$$\Sigma_{ee} \sim \Sigma_{SC}(kT/\eta)^2 \tag{5.2}$$

That is, at room temperature in a typical metal, Σ_{SC} is reduced from its small intrinsic value by a factor $\sim 10^4$.

As a consequence of the form of (5.2), it is clear that any electrical resistivity ρ_{ee} arising from electron–electron scattering must vary as T^2. It is, therefore, generally accepted that the criterion for the existence of such scattering is the presence of a T^2 term in the low-temperature resistivity. Nevertheless, in view of the preceding paragraph, it is not surprising that for all "simple" monovalent* metals such a resistivity has never been observed.

However, for transition metals the situation is completely different. Components of resistivity varying as T^2 have been seen at low temperatures in no fewer than 18 of the 24 elements [71V1], and it is clear, as first pointed out by Baber [37B1], that some form of electron–electron scattering is most important. The qualitative argument is that for (s, d) to (s', d') collisions the comparatively light, mobile s-electrons are strongly scattered on colliding with the heavy d-electrons, and the factor $(kT/\eta)^2$ is modified to $(kT)^2/\eta(\varepsilon_d - \eta)$, where ε_d is the upper edge of the d-band. As $(\varepsilon_d - \eta)$ can be as much as an order of magnitude smaller than η, the degeneracy factor is therefore considerably increased. Mott [64M1] has pointed out that for a collision in which the s-electron makes a *transition* to the d-like part of the Fermi surface—that is, an (s, d) to (d', d'') process—this factor is still further increased to $(kT)^2/(\varepsilon_d - \eta)^2$. Colquitt *et al.* [71C1], in a paper on thermopower arising from electron–electron scattering that we shall return to later, also conclude that s to d transitions are the most effective electron–electron scattering mechanism.

5.1c   *Magnetic Effects: Collective Electrons and Isolated Spins*

Any survey of transport properties in transition metals and alloys must include some account of the complications arising from magnetic effects. For example, even among the elements, some are *ferromagnetic* (Co, Fe, Ni), some are *antiferromagnetic* (Mn, Cr), while others are *"almost"* *magnetic*

---

* A T^2 component of resistivity has been observed in Al and In below 4 K by Garland and Bowers [68G2], and later attributed by Lawrence and Wilkins [72L3] to electron–phonon Umklapp scattering.

(e.g., Pd, Pt), with magnetic susceptibilities that are greatly enhanced from the values deduced from specific heat measurements.

The position regarding magnetic interactions is further complicated by the widespread use of several different models, of which some, but not all, are equivalent. Although on all such models the magnetic effects are attributed to an interaction between d-electrons, the detailed mechanisms can be very different.

One such model is the *band* or *collective electron* model first introduced by Stoner [38S1]. The two essential features of this model are: (a) The d-electrons form relatively wide energy bands and are mobile (itinerant) throughout the metal. (b) The d-band is split by the exchange energy into spin-up and spin-down components, separated in energy as shown in Figure 5.2. The spontaneous magnetization per atom at the absolute zero is then simply given by the Bohr magneton times the difference per atom of the occupation of the two sub-d-bands.

As far as transport properties are concerned, it is clear that they can be discussed by using an extension of the Mott s-d model as outlined in Section 5.1a, with the essential difference that we now have to contend with two directions of spin. Equation (5.1) must be extended to allow for two parallel currents of opposite spin, $\sigma(\uparrow)$ and $\sigma(\downarrow)$, that are given by

$$\sigma(\uparrow) = \sigma_s(\uparrow) + \sigma_d(\uparrow)$$
$$\sigma(\downarrow) = \sigma_s(\downarrow) + \sigma_d(\downarrow)$$

(5.3)

Fortunately, at very low temperatures in the impurity (residual) resistance region, it is reasonable to assume that there is no change of spin at each collision and that the two parts of (5.3) are independent. This is the basis of the *two-current* combination model that has recently been used extensively in the study of ferromagnetic metals (see, for example [69F1, 70D2, 73P1]). On this basis the study of ferromagnetic metals is comparatively simple—at least in principle—and this is particularly so in the case of nickel. For that metal near $T = 0$ K the $d\uparrow$ band is full, so that only $s\downarrow$ electrons can make s-d transitions. The total conductivity is then largely determined by the $\sigma(\uparrow)$ component, and to this extent, conduction in nickel in the impurity-scattering regime is more like that of a noble metal than a transition metal.

At higher temperatures, however, the situation becomes more complicated for two reasons, both of which we shall develop in later sections. These are: (a) For $T > 0$ electrons are increasingly subject to spin-flip scattering between the two sub-bands in a process known as *spin mixing*, so that the two groups of electrons can no longer be regarded as independent. (b) With

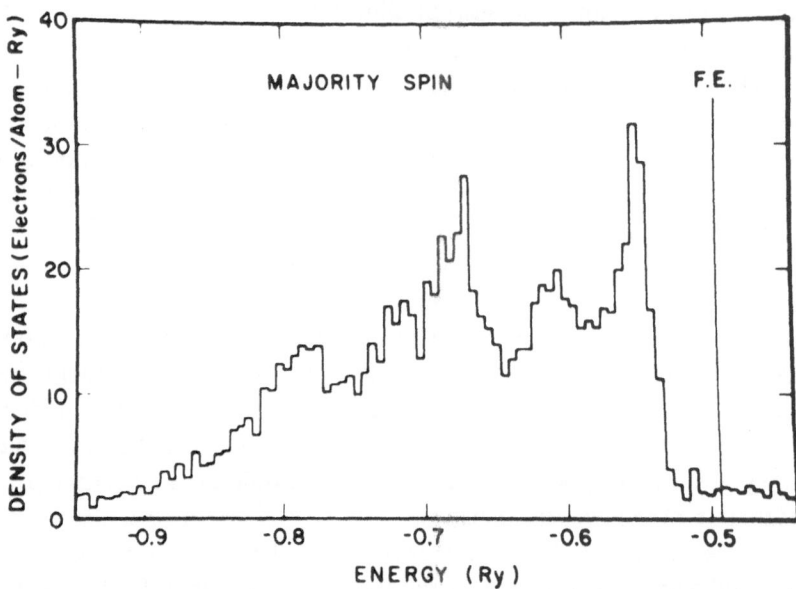

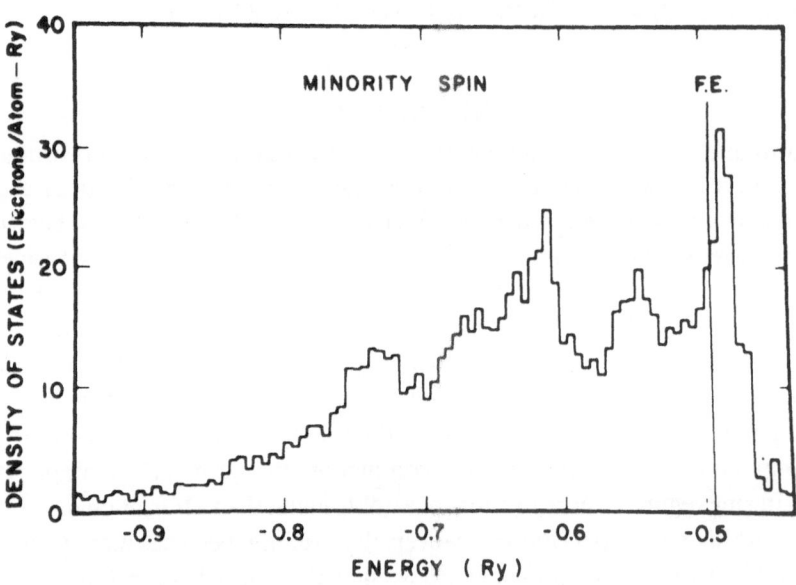

Fig. 5.2. The calculated density of states in nickel at $T = 0$ K, showing the effect of spin splitting (from [72L1]).

increasing T, the energy gap between the two components of the split d-band diminishes and finally disappears at the Curie temperature.

As an alternative to the collective electron model there is the *localized-spin* model, according to which the unpaired "magnetic" electrons are localized on particular atoms. Exchange interactions then take place between the conduction s-electrons and the localized d-electrons, so that s-electrons of opposite spin see a difference in energy at a scattering ion of $2J\mathbf{S} \cdot \mathbf{s}$, where $\mathbf{S}$ and $\mathbf{s}$ are the spin operators of the ion and conduction electron, respectively, and where J is the Heisenberg exchange integral. The resultant additional resistivity $\rho_{\text{spin-dis}}$ is said to be due to *spin-disorder* scattering, with a value that increases from 0 at $T = 0$, where all the spins are aligned, to a constant above the Curie (or Néel) point, where the spins are completely disordered.

Many authors have attempted to provide evidence from resistivity measurements favoring one or the other of these two models. Mott and Stevens [57M2] and Coles [58C3] have drawn attention to the contrasting behavior of the temperature dependence of ρ above the Curie points of iron and nickel. In the former metal ρ is very much less dependent on T than in the latter, suggesting, at least at high temperatures, a spin-disorder picture for iron and a collective-electron model for nickel. Schwerer and Cuddy [70S1] have used a similar argument to conclude that the spin-disorder model gives the best description of the magnetic resistivity in *alloys* based on *both* iron and nickel. However, at low temperatures the experimental evidence seems to suggest that the two models are more complementary. In this case, part of the evidence is centered on the study of *magnons*.

5.1d *Magnetic Effects: Magnons and Paramagnons*

The breakdown of perfect spin alignment above $T = 0$ K can be treated in terms of spin-wave excitations, and at low temperatures the number of noninteracting magnons is proportional to $T^{3/2}$ [71K1]. Consequently, at these temperatures $\rho_{\text{spin-dis}}$ might also be expected to vary as $T^{3/2}$, and this temperature dependence is observed experimentally in dilute alloys, such as Pd–Fe [69W1] and Ni–Mn [71M1] alloys in which the spins are spatially disordered. Long and Turner [70L1] and Kagan *et al.* [68K1] have confirmed in a detailed theoretical analysis that this is the correct form of $\rho_{\text{spin-dis}}$ for these materials.

On the other hand, for *pure* ferromagnetic transition metals the situation is likely to be quite different. There the spins are spatially periodic, and the scattering by magnons is likely to be coherent rather than incoherent. Many authors have shown that in such circumstances the low-temperature "magnon" resistivity varies as T^2 (see, for example, [56K1, 63G2, 71M1]), and this is indeed the temperature dependence observed experimentally. (It is also interesting to note that this is the temperature dependence in P̲d–Fe containing over 1% Fe [70S2].)

Obviously it is rather confusing that the temperature dependence predicted for electron–magnon scattering in pure magnetic materials is identical to that for electron–electron scattering, and this has aroused considerable controversy in the study of magnetic transition metals. However, we again draw attention to the fact that a resistivity varying as T^2 is observed in the majority of transition metals, nonmagnetic as well as magnetic, and we suggest that the generality of this observation confirms the importance of electron–electron scattering.

A different complication arises in the study of metals, such as palladium and platinum, that are "almost" ferromagnetic. For these metals the exchange interaction, although not quite strong enough to cause large-scale spin alignment, is sufficient to produce short-lived local spin fluctuations that can again act as scattering centers for the conduction electrons [66B1, 66D1]. The quanta of these critically damped spin waves are known as *paramagnons*. Mills and Lederer [66M1] and Schindler and Rice [67S1] have shown that (as might be expected) electron–paramagnon scattering at very low temperatures also results in a resistivity varying as T^2, and they therefore attribute the observed temperature dependence in palladium and platinum and their alloys to that source. Their interpretation is strongly supported by measurements on the P̲d–Ni system. In these alloys both the magnetic susceptibility and the coefficient of T^2 in the electrical resistivity rise rapidly with increasing nickel concentration. Lederer and Mills [68L1] show how these observations can be interpreted on the basis of local spin-density fluctuations at the impurity (nickel) sites additional to the uniform spin fluctuations in the palladium host. Nevertheless, it is interesting to note that Lederer and Mills point out that the T^2 temperature dependence "is obtained because the scattering mechanism is an electron–electron process; the s-electrons scatter inelastically against the local d-electron spin fluctuations. . . ."

Furthermore, Kaiser and Doniach [70K2] have shown that at rather higher temperatures where $T \sim T_{sf}$, the characteristic temperature of the spin fluctuations, the resultant resistivity is proportional to T. The reason is that then the paramagnons behave as quasi-bosons, giving a linear resistivity that is the exact analogue of the electron–phonon resistivity at $T \sim \theta_D$. The various models are perhaps much more similar than they appear at first sight.

5.1e Magnetic Effects: Spin Mixing

In Section 5.1c we mentioned briefly the spin-mixing model first introduced by Gomes [66G2] and Campbell et al. [67C1]. In ferromagnetic metals and alloys at low temperature, electrons of opposite spin are independent, with impurity resistivities in the two spin directions of $\rho_0(\uparrow)$ and $\rho_0(\downarrow)$. Consequently, the net impurity resistivity—the observed residual resistivity ρ_0^{LT}—is given by

$$\rho_0^{LT} = \frac{\rho_0(\uparrow)\rho_0(\downarrow)}{\rho_0(\uparrow) + \rho_0(\downarrow)} \tag{5.4}$$

At high temperatures, on the other hand, as the result of electron–electron or electron–magnon collisions, electrons are continually flipped (or "mixed") between the two spin directions with a time between collisions that we label $\tau(\uparrow\downarrow)$. With increasing temperature $\tau(\uparrow\downarrow)$ becomes very much less than either of the relaxation times, $\tau_0(\uparrow)$ and $\tau_0(\downarrow)$, for impurity scattering. In the high-temperature limit, electrons will therefore have an equal probability of having spin $\uparrow$ or $\downarrow$; so, for any one spin direction, the impurity resistivity will be given by the average value of $\frac{1}{2}[\rho_0(\uparrow) + \rho_0(\downarrow)]$. The net high-temperature impurity resistivity is, therefore,

$$\rho_0^{HT} = \frac{1}{4}[\rho_0(\uparrow) + \rho_0(\downarrow)] \tag{5.5}$$

Unless $\rho_0(\uparrow) = \rho_0(\downarrow)$—as in a nonmagnetic metal—a little algebra shows that $\rho_0^{HT} > \rho_0^{LT}$, in which case there exist deviations from Matthiessen's rule [68F1]. Furthermore, it is apparent that the greater the ratio $\rho_0(\uparrow)/\rho_0(\downarrow)$, the greater the deviations. Indeed, it was in order to explain the unusually large deviations observed in alloys based on Ni, Co, and Fe (for a summary, see [71F1]) that the model was introduced in the first place, although Greig and Rowlands [74G4] have recently shown that it is possible to explain these anomalies in other ways.

Although the model has gained fairly widespread acceptance, there is still some doubt about the origin of the mixing mechanism. The choice between electron–electron and electron–magnon scattering has not been fully resolved. In a physical sense it is perhaps easiest to attribute the mixing to electron–magnon processes. Fert [69F1] has given a theoretical analysis supporting this view; unfortunately this appears to lead to a paradox. If nonresistive $s\uparrow$ to $s\downarrow$ magnon scattering is sufficiently great to lead to reasonable values of $\tau(\uparrow\downarrow)$, then clearly *resistive* scattering transitions from $s\uparrow$ to $d\downarrow$ are even more likely, and numerical estimates show that these would lead to an electron–magnon T^2 resistivity more than an order of magnitude greater than that actually observed [68F1]. For this reason, it is perhaps more logical to attribute the phenomenon to electron–electron N-type collisions, because these have less correspondence with s–d resistive processes.

5.1f *Magnetic Effects: the Curie and Néel Temperatures*

With increasing temperature, ferromagnetic metals and alloys gradually lose their spontaneous magnetization, becoming completely paramagnetic above the Curie temperature T_C. It has long been recognized that the curves representing the temperature dependence of resistivity and thermopower undergo a change of slope at T_C, and this has been explained—qualitatively at least—on both the split-band and spin-disorder models outlined in previous sections. Indeed, one of Mott's earliest papers on s–d scattering [36M1] gave a remarkably good account of the broad features of the variation of both ρ and S with T in nickel, including the changes at T_C.

In the last few years, however, interest has centered on the behavior of ρ and S within a few degrees of T_C. The reason is that the magnetic phase transition is now recognized as a critical phenomenon [67F1], with the specific heat C exhibiting a lambda anomaly—that is, divergent behavior—at the critical point T_C [67H1]. In addition, it has also been shown that *both* $d\rho/dT$ [67C2] and dS/dT [74T3, 74P1] exhibit lambda peaks at T_C (Figure 5.3). In fact, in the neighborhood of T_C the temperature variation of C, $d\rho/dT$, and dS/dT can all be represented by the same type of (divergent) empirical law. For example, in the case of resistivity, the law is

$$\left(\frac{1}{\rho_C}\right)\left(\frac{d\rho}{dT}\right) = \left(\frac{A}{\lambda}\right)(\delta^{-\lambda} - 1) + B \tag{5.6}$$

where $\delta = |(T - T_C)|/T_C$, ρ_C is the resistivity at the Curie point, and A, B, and λ are constants. In general, data are best for $T > T_C$, and it has been shown that for pure Ni, λ^+ (i.e., λ for $T > T_C$) is ~0.1 for each of the three variables C, $d\rho/dT$, and dS/dT. This divergence of the derivative has been explained by Fisher and Langer [68F2] on the basis of scattering from

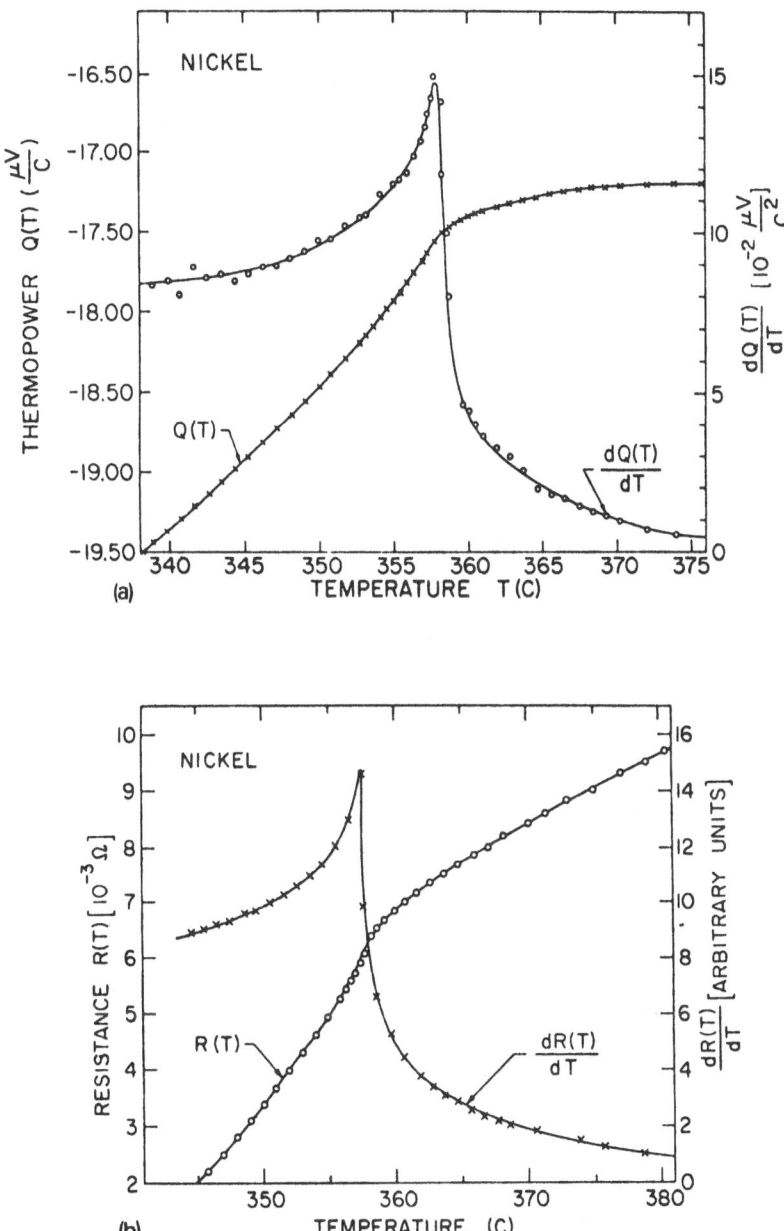

Fig. 5.3. Thermoelectric power (a), here symbolized by $Q(T)$, and resistivity (b) of nickel near the Curie temperature (from [74T3]).

short-range spin fluctuations. However, the extension of this work to thermopower is controversial, and we shall return to this in a later section.

Although the three best-known "magnetic" transition metals, Fe, Co, and Ni, are ferromagnetic, two further elements have ordered magnetic structures at low temperatures. These are the antiferromagnetic metals Cr and Mn, with Néel temperatures T_N of 312 K and 94 K, respectively. Once again, the thermopower is observed to change rapidly with T in the neighborhood of T_N. As with ferromagnetism, antiferromagnetism can be discussed either on the basis of localized or itinerant models, although for Cr—and possibly Mn—the evidence rather strongly favors the latter [66H1]. The suggestion is that the magnetic structure can be described in terms of spin-density waves, with a periodic modulation, as deduced from neutron diffraction measurements, of more than 20 times the length of the unit cell. When the temperature is lowered through T_N, the Fermi surface of such a metal is considerably modified, with subsequent changes in the values of ρ and S.

5.2   The Diffusion Thermopower of Transition Metals

5.2a   Phonons and Impurities

The net diffusion thermopower of a two-band conductor in which the carriers are s- and d-electrons is

$$S_d = \frac{\sigma_s S_d^s + \sigma_d S_d^d}{\sigma_s + \sigma_d} \tag{5.7}$$

where S_d^s and S_d^d are, respectively, the thermopowers of the s- and d-electrons when either group is acting independently of the other. In the original Mott model it was assumed (a) that $\sigma_s \gg \sigma_d$ and (b) that S_d^d was not anomalously large. On this basis it is clear that $S_d \sim S_d^s$, and this is the approximation used in virtually all attempts at comparing theory and experiment.

The second of these assumptions can be easily justified if we assume that $\sigma_d \propto N_d(\varepsilon)\tau_{dd}$, where τ_{dd} is the relaxation time for d–d scattering. Then, as $\tau_{dd} \propto 1/N_d(\varepsilon)$, it follows that σ_d is independent of $N_d(\varepsilon)$ and $S_d^d \propto$ $(d \ln \sigma_d/d\varepsilon)_{\varepsilon=\eta}$ does not contain the term $[d \ln N_d(\varepsilon)/d\varepsilon]_{\varepsilon=\eta}$ which, in transition metals, can be very large. On the other hand, since the conductivity of the s-electrons is limited by s–d scattering, we see that $\sigma_s \propto N_s(\varepsilon)\tau_{sd}$, and hence $\sigma_s \propto N_s(\varepsilon)/N_d(\varepsilon)$. Consequently, the s-component of diffusion thermopower is dominated by the term $[d \ln N_d(\varepsilon)/d\varepsilon]_{\varepsilon=\eta}$, so that, to a first approximation

$$S_d \sim S_d^s \sim \frac{\pi^2 k^2 T}{3|e|} \left(\frac{d \ln N_d(\varepsilon)}{d\varepsilon} \right)_{\varepsilon=\eta} = \frac{\pi^2 k^2 T}{3|e|} \left(\frac{1}{N_d(\varepsilon)} \frac{dN_d(\varepsilon)}{d\varepsilon} \right)_{\varepsilon=\eta} \quad (5.8)$$

It is clear from this equation why the large values of $dN_d(\varepsilon)/d\varepsilon$ found in many transition metals lead directly to large values of S_d. The complete equation for S_d^s in a transition metal is

$$S_d^s = -\frac{\pi^2 k^2 T}{3|e|} \left(\frac{3}{2\eta} - \frac{1}{N_d(\varepsilon)} \frac{dN_d(\varepsilon)}{d\varepsilon} \right)_{\varepsilon=\eta} \quad (5.9)$$

where, as discussed in Chapter 2, the term $3/2\eta$ arises from taking the logarithmic derivative of electron velocity and Fermi surface area in the approximation of spherical energy surfaces. Consequently, except when $[dN_d(\varepsilon)/d\varepsilon]_{\varepsilon=\eta} \sim 0$, Eq. (5.9) can be approximated by Eq. (5.8).

Although this model was first proposed by Mott in 1936, it still gives the best understanding of thermopower in transition metals, and, in recent years, has formed the basis of a number of detailed calculations. Kolomoets [66K1] has considered interband scattering in the case of two overlapping parabolic bands, consisting of a wide shallow band of "light" carriers spanning a high narrow band of "heavy" carriers. His calculated S_d is given as a function of $\gamma = (\eta - \varepsilon_d)/kT$, where ε_d is the d-band edge closest to η, and the variation is shown in Figure 5.4 for heavy bands that are almost empty (case 1), and almost full (case 2). There are two important conclusions to be drawn from this diagram: (a) The diffusion thermopower is of opposite sign in metals in which the "heavy" d-band is almost full to those in which it is almost empty. (b) In both cases the value of S_d is a maximum when $\gamma \sim 0$; that is, when the Fermi level lies close to the band edge. Both of those features can be understood qualitatively if S_d is dominated by a term $[d \ln N_d(\varepsilon)/d\varepsilon]_{\varepsilon=\eta}$.

The experimental variation of S with T, mainly between 80 and 1800 K, has been reviewed recently by Vedernikov [69V1]. His diagram summarizing the results for all elements except technetium is reproduced in Figure 5.5. We shall discuss some of the details of this diagram in later sections, but for the present we merely point out that, on the whole, the thermopower at the highest temperatures is positive for elements from groups III to VI and negative for those from the three divisions of group VIII; that is, positive

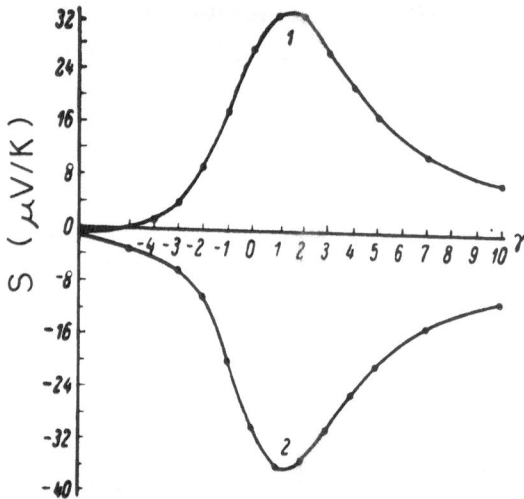

Fig. 5.4. Calculated dependence of the thermoelectric power on the parameter $\gamma = (\eta - \varepsilon_d)/kT$, (1) for d-bands almost empty and (2) for d-bands almost full (from [66K1]).

when the d-band is largely empty but becoming more negative as it fills in accordance with Kolomoets' predictions.

More recently, further measurements have been made of the variation of S with T in eleven transition elements up to ~ 1300 K by Nemchenko and his co-workers [70N3, 72N1]. Their results are very similar to those reported by Vedernikov, but the main feature of their work is that they have used the measurements in conjunction with Eq. (5.9) to make qualitative predictions about the energy dependence of $N_d(\varepsilon)$ close to the Fermi level. Their conclusions for the elements Re, Os, Ir, and Pt are reproduced in Figure 5.6.

Another, more ambitious calculation along similar lines applied specifically to the pure metals Mo, W, Rh, Ir, Pd, and Pt has been carried out by Aisaka and Shimizu [70A1]. These authors make use of de Haas–van Alphen data to give the densities and effective masses of the electrons and holes (one group of each in Pd and Pt and two groups of each in Mo, W, Rh, and Ir), the specific heat data to give the densities of states at η, and the rigid band model to determine the variation with T of the mean densities of states at η. For Mo, W, Pd, and Pt the qualitative agreement between the observed and calculated variation of S with T is excellent, although the absolute

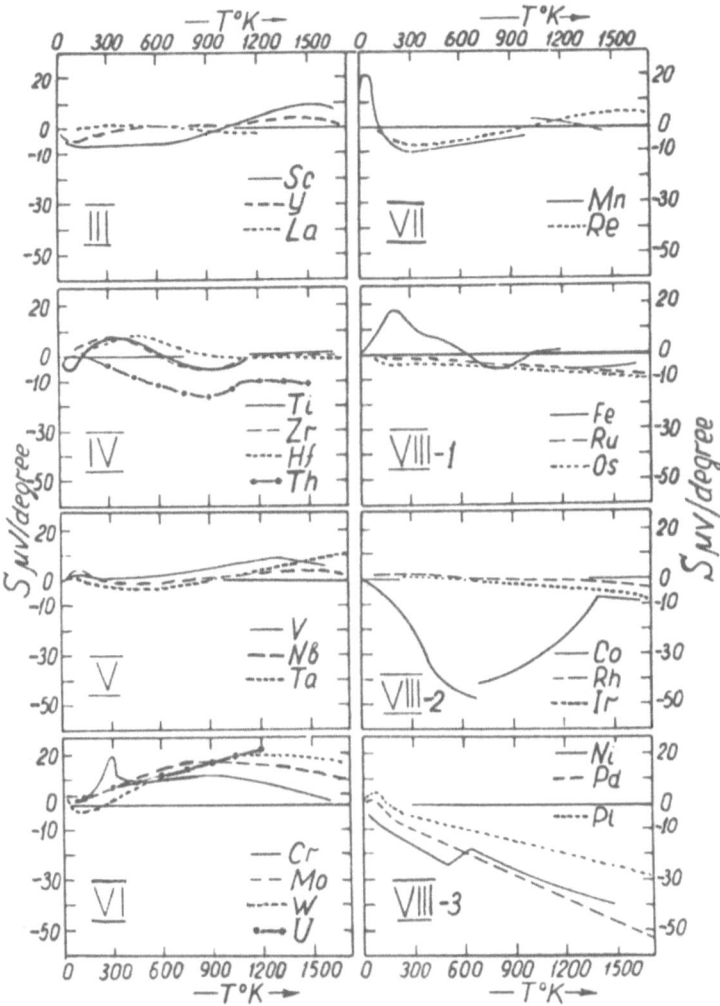

Fig. 5.5. The thermoelectric power of transition metals (from [69V1]). Roman numerals refer to the column in the periodic table.

magnitude of the theoretical curves is too great by about a factor of 2. For Rh and Ir the agreement is significantly worse.

These calculations were aimed at deriving S_d at room temperature and above, on the assumption that s–d collisions are entirely the result of electron–phonon scattering. Aisaka and Shimizu have described the phonons by the simplest possible Debye model with matrix elements averaged

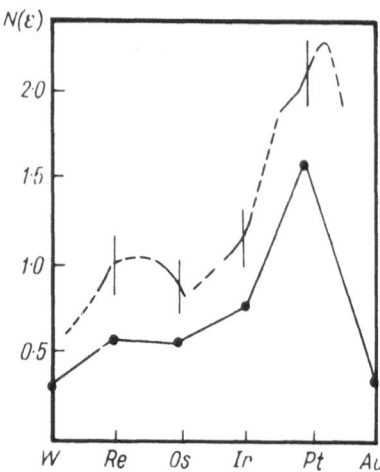

Fig. 5.6. Comparison of the density of states of W, Re, Os, Ir, Pt and Au as deduced from specific heat, solid line, and from an analysis of the temperature dependence of the thermoelectric power, dashed line (from [72N1]).

over the surfaces of equal energy in the Brillouin zone. In the conclusion to their paper, they point out that the discrepancies between theory and experiment may well be attributable to these simplifications. They also point out that the discrepancies could result from the neglect of electron–electron and electron–paramagnon collisions, and we shall enlarge on this aspect of the problem in later sections.

For the present, we should like to turn to similar calculations of S_d^s based on the Mott model for s–d collisions at *impurities*. Examples of these calculations have been given by Fletcher and Greig [65F1, 68F3] and Dugdale and Guénault [66D2]. Using essentially Eq. (5.8), the former authors found that, for Pd–Ag alloys containing between 1% and 20% Ag, there was remarkably good agreement between theory and experiment when the term $[d \ln N_d(\varepsilon)/d\varepsilon]_{\varepsilon=\eta}$ was obtained from the rigid band model in conjunction with specific heat data. For a rigid band model, $N_d(\varepsilon)$ is unchanged on alloying, and $N_d(\varepsilon)d\varepsilon = Z\,dc$, where c is the atomic fraction of alloying element of valence difference Z from the host. The factor Z is required to give the direction of movement of the Fermi energy on alloying. For example, when Ag is added to Pd, $Z = +1$, the average number of electrons in the d-band is increased, and in comparison to pure Pd, η changes to a higher value. Thus $d \ln N_d(\varepsilon)/d\varepsilon = [1/N_d(\varepsilon)][dN_d(\varepsilon)/d\varepsilon] \propto d\gamma/N_d(\varepsilon)d\varepsilon \propto (1/Z)(d\gamma/dc)$, where γ is the measured electronic specific heat divided by T. As a typical example of the surprisingly good agreement with experiment obtained with this simple model, we compare the calculated result in Pd + 10% Ag of $S_d^s/T = -0.131$ μV deg^{-2} with the experimental value at helium temperatures of $-0.13\ \mu V\ deg^{-2}$. Similar excellent agreement was obtained in Pt–Au alloys.

A further interesting feature of this analysis that if Rh ($Z = -1$) is added to Pd and electrons *removed* from the d-band, the position of η will move toward and through the peak of highest energy in the density of d-states (Figure 5.1). It therefore follows from the above analysis that we expect S to change sign at the same Rh concentration as $d\gamma/dc$ changes sign ($\sim$4%), and this is precisely what is observed. A somewhat more sophisticated version of this model was developed by Dugdale and Guénault [66D2] to account for the concentration dependence of resistivity and thermopower across the complete composition range of the Pd–Ag system [74G7].

Unfortunately, this simple analysis which is so successful in alloys in which Pd and Pt are alloyed with their nearest neighbors in the periodic table has hardly been tested elsewhere. Until further experiments are carried out, it certainly must be used with caution, particularly since it has been established that in certain other Pd alloys—for example, Pd–Pt, Pd–V, Pd–Mo, and Pd–Ru [72B3, 67A1, 72G1]—the simple calculations outlined above break down. Nevertheless, it is worth remarking that for all Pd alloys in which specific heat data are available—and here we include Pd–Ru in which the rigid band model is inapplicable—there is a remarkable experimental correlation between the sign and magnitude of the measured low-temperature S and the sign and magnitude of $d\gamma/dc$ [72G1]. Brereton [72B4] has explained the Pd–Ru results by including in the calculation the change of shape of the density of states curve resulting from impurity-scattering effects, and it appears that this more sophisticated approach is necessary when $|Z| > 1$.

5.2b *Electron–Electron Scattering*

Although electron–electron effects are crucial to an understanding of electrical resistivity in transition metals, their influence in determining thermopower has been almost completely neglected. Part of the reason for this is that most detailed studies have been concerned with data either taken around room temperature and interpreted in terms of electron–phonon scattering or measured at very low temperatures where electrons are predominantly scattered by impurities. In addition, because of the large number of parameters involved, it is difficult to compare theory and experiment in an unambiguous way.

The only detailed study of the topic has been made by Colquitt *et al.* [71C1]. Their analysis was based on the usual Mott *s–d* scattering model but with additional assumptions about the curvatures of the *s*- and *d*-bands and the relative magnitudes of the Fermi wave vectors k_s and k_d. It was found that the sign as well as the magnitude of S_d is very sensitive to these parameters. Colquitt *et al.* emphasize that their work must be regarded as something of a model calculation, because all the finer details of the Fermi surface are completely neglected. Nevertheless, within these rather severe limitations, they show that the diffusion thermopower S_d^{ee} arising from electron–electron effects can make a substantial contribution to the total diffusion thermopower, both at very low temperatures $(T \sim 0.03\theta_D)$ and also at much higher temperatures $(T > \theta_D)$.

In order to explain how these electron–electron effects can be distinguished, it is important to emphasize that Colquitt *et al.* show that (a) as with diffusion thermopower in general, $S_d^{ee} \propto T$, and (b) the magnitude of S_d^{ee} lies in the same range as S_d from other forms of electron scattering. There is, therefore, nothing "unusual" about S_d^{ee}, and its unambiguous separation from the measured S is rather difficult. The relative importance of S_d^{ee} in comparison to other components of S_d is determined through the weighting formula

$$S_d = \frac{\Sigma_j W_j S_d^j}{\Sigma_j W_j} \qquad (5.10)$$

where S_d^j and W_j are, respectively, components of diffusion thermopower and thermal resistivity arising from the *j*th scattering mechanism. Thus, it is clear that when any one scattering mechanism dominates, then S_d is simply proportional to T. Therefore, anomalies in S_d are only observed when there is a change with temperature in the *relative importance* of two or more scattering processes. It is for this reason that at very low temperatures there is a predicted change in slope when the dominant mechanism changes, with increasing T, from electron–electron to electron–phonon, and again at much higher temperatures $(T > \theta_D)$ when electron–electron processes reassert themselves.

In this latter case Colquitt *et al.* suggest that there is a certain temperature range in which the electronic thermal resistivity W_{e-ph} (independent of T) is much greater than the electron–electron resistivity W_{e-e} $(\propto T)$, although the products $W_{e-ph}S_d^{e-ph}$ $(\propto T)$ and $W_{e-e}S_d^{ee}$ $(\propto T \times T)$ are comparable. Consequently, the total S_d as given by (5.10) should vary as $(\alpha T +$

βT^2), and the authors suggest that this is supported by the observed temperature dependence in pure Mo and W above θ_D. Unfortunately, we recall that Aisaka and Shimizu [70A1] explained this temperature dependence equally well on the basis of electron–phonon scattering only. Trodahl [73T1] has shown that the measured S of very pure tungsten at very low temperatures (<5 K) *could* result from the combination of a negative electron–impurity thermopower and a positive S_d^{ee}. However, the difference in fit between this and the simple addition of a linear diffusion term and a cubic phonon-drag thermopower is very slight. More recently Garland and Van Harlingen [74G2], on the basis of measurements of the thermoelectric ratio G, the resistivity, and thermal conductivity, have questioned the relevance of electron–electron scattering in tungsten at low temperatures.

5.2c *Magnons and Paramagnons*

The problem of thermoelectricity in ferromagnetic metals at low temperatures was first discussed by Kasuya [59K1]. In his paper—which was an extension of earlier work on the resistivity of ferromagnetic metals [56K1]—he adopted a localized-spin model for the d-electrons and concluded that under certain favorable circumstances the thermopower of magnetic materials could be extremely large.

The reason for this is the coexistence of asymmetric elastic scattering (explained below) and inelastic scattering at the magnetic ions, and it is this *combination* of events that leads to the anomalous behavior. This problem of simultaneous elastic and inelastic scattering has been discussed in rather more general terms by Guénault and MacDonald [61G2], and they showed why such circumstances can lead to a thermopower as great as the classical limit of $\pm k/e$ ($\pm 86 \mu$V/K). The point is that S_d depends on the energy dependence of the conductivity and this, for most scattering processes, is not an extreme variation, with $(d \ln \sigma / d \ln \varepsilon)_{\varepsilon = \eta} \sim 1$. If, however, there is an inelastic process present so that one group of electrons—say, electrons of one spin direction—absorb energy on collision, while the remainder—the electrons of opposite spin—emit energy, the two groups become respectively displaced above and below η. The energy variation of conductivity is now much more extreme, but as long as the groups are symmetrically displaced about η, the effect on S_d is negligible. However, if at the same time one group is elastically scattered much more than the other, the distribution about η becomes highly asymmetrical, and in the extreme limit, where the two components of σ, σ_1 and σ_2, are separated in energy by $\sim kT$,

$$\left(\frac{1}{\sigma}\frac{d\sigma}{d\varepsilon}\right)_{\varepsilon=\eta} \sim \frac{1}{\sigma_1}\left(\frac{\sigma_1-0}{kT}\right) \sim \frac{1}{kT} \qquad (5.11)$$

that is to say, $S_d \sim k/e$.

It is worth emphasizing that this value of S_d, achieved by separating the two components of σ in energy, can only be obtained in rather special circumstances. If the inelastic energy Δ is much less than kT, then σ_1 and σ_2 are not clearly separated. Similarly, if $\Delta \gg kT$ the separation cannot be achieved at all. In addition, it is essential that the two relaxation times for elastic scattering be different or $d\sigma/d\varepsilon$ simply averages to zero. Although the chances may be small that all these conditions are optimized simultaneously, it is clear that conditions in ferromagnetic metals are exactly as required for at least partial observation of the effect.

Kasuya [59K1] considered a localized-spin model in which the inelastic processes are due to energy absorption/emission at the ionic levels generated by an internal molecular field, but in which the elastic spin-disorder scattering is also different for electrons of opposite spin. That is, conditions are such that a large anomalous thermopower is expected. However, Kasuya did not attempt a detailed comparison with experiment. A rather similar analysis was given independently by de Vroomen and Potters [61D2], although their work was mainly concerned with the problem of giant thermopowers in "Kondo" alloys.

More recently, Korenblit and Lazarenko [71K2] have again examined these ideas but have applied them to the band model of ferromagnetism with inelastic processes attributed to electron–magnon collisions. For example, in the case of nickel near $T = 0$ K, spin-$\uparrow$ electrons can only absorb magnons, while spin-$\downarrow$ electrons can only create them. At the same time, in many nickel alloys elastic collisions at the solute atoms are completely different for the two directions of spin. So, on this model also we have precisely the condition necessary for the existence of thermopower anomalies.

Korenblit and Lazarenko presented in their paper a detailed theory of the effect, and their analysis leads to three major predictions: (a) The observed S_d should be proportional to the difference in the impurity relaxation times $[\tau_0(\uparrow) - \tau_0(\downarrow)]$, so that, for different alloys based on one parent metal, the sign of S_d should depend on whether the ratio $\tau_0(\uparrow)/\tau_0(\downarrow)$ is greater or less than unity. (b) At very low temperatures, S_d should first increase exponentially due to the energy gap restricting single-magnon excitations and then continue to rise roughly as T, reaching a maximum

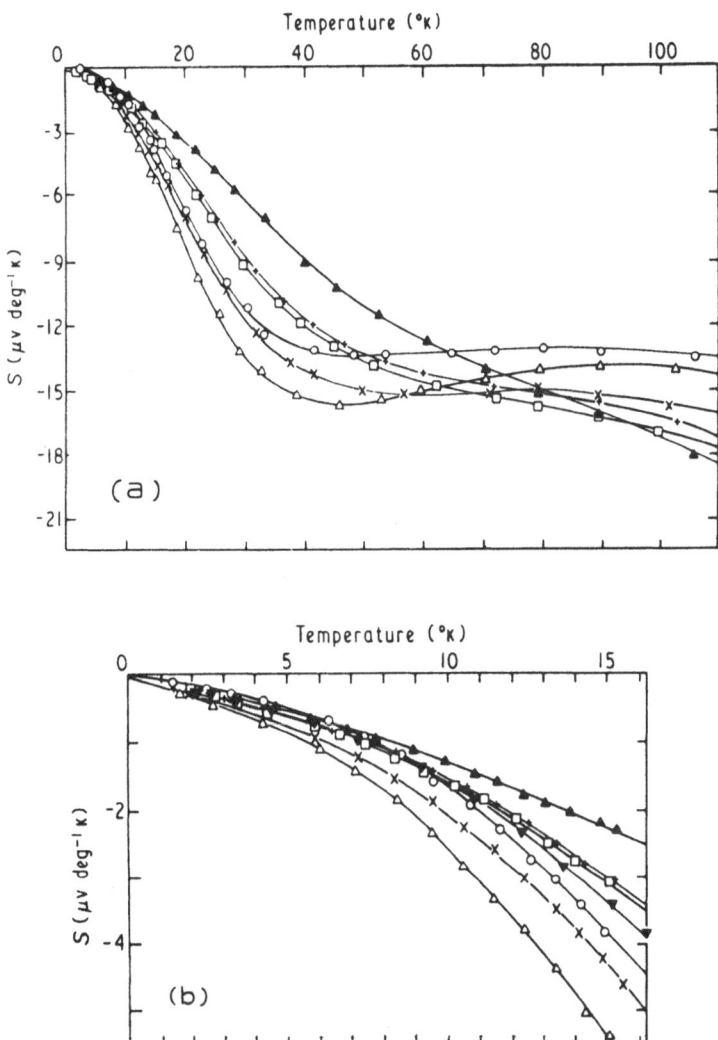

Fig. 5.7. (a) The temperature dependence of the absolute thermoelectric power of Ni–Co and Ni–Fe alloys. The symbols represent △ Ni + 1 at.% Co; ×Ni + 2 at.% Co; + Ni + 5 at.% Co; ○ Ni + 0.5 at.% Fe; □ Ni + 2 at.% Fe; ▲ Ni + 5 at.% Fe. (b) shows the low temperature data in detail (from [70F2]).

when the relaxation time for electron–magnon scattering is approximately equal to $\sqrt{\tau_0(\uparrow)\tau_0(\downarrow)}$. At higher temperatures, electron–phonon scattering (and spin mixing) become increasingly important, and the effect is washed out. For dilute nickel alloys the temperature of these maxima is estimated to be in the region of 20 to 60 K. (c) The value of S_d at the maximum is inversely proportional to the solute concentration.

In comparing these predictions with experiment, Korenblit and Lazarenko used the measurements of Farrell and Greig [70F2] on dilute Ni–Cu, Ni–Co, and Ni–Fe alloys. The data for the latter two series are reproduced in Figure 5.7. It is clear that the observations agree qualitatively with predictions (b) and (c), and Korenblit and Lazarenko found that the parameters required to fit their theoretical expressions to the data were all reasonable. Furthermore, we find that in Ni–Cr, for which $\tau_0(\uparrow)/\tau_0(\downarrow) < 1$ in comparison to the other three alloys mentioned for which $\tau_0(\uparrow)/\tau_0(\downarrow) > 1$, the sign of S_d is anomalously positive; that is, prediction (a) is also confirmed. Nevertheless, it must be pointed out that, in spite of the apparent success of the model, it is almost impossible to differentiate these maxima from phonon-drag peaks. All we can add is that it would be somewhat surprising if, for all dilute ferromagnetic metals, phonon drag were completely absent.

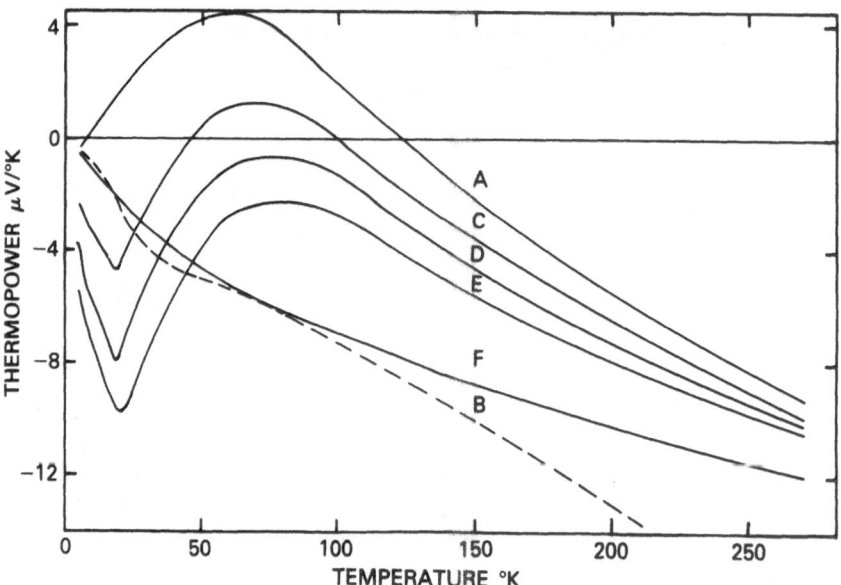

Fig. 5.8. Thermoelectric power of Pd–Ni alloys, Curve A is the pure Pd data of Fletcher and Greig [65F1] and Taylor and Coles [56T1]; curve B is the annealed pure Ni data of Blatt *et al.* [67B4]; curves C, D, E, and F are the data for Pd–Ni alloys with the at.% of Ni being as follows: C, 0.5; D, 1.0: E, 1.66 and F, 6.16 (from [68F4]).

With regard to materials in which the scattering is predominantly by paramagnons, comparatively little work has been done. The only thermoelectric measurements of note have been made on Pd–Ni alloys by Foiles and Schindler [68F4], on Ir–Fe alloys by Sarachik and Touger [74S5], and on Rh–Fe alloys by Nagasawa [68N2] and by Graebner et al. [74G9]. In dilute Pd–Ni alloys the T^2 resistivity rises rapidly with increasing Ni content, and this is attributed to an increase in the spin-density fluctuations at the nickel sites [68L1]. Foiles and Schindler's measurements are shown in Figure 5.8, and there are two points of special interest. The first is that in the dilute *paramagnetic* alloys there are pronounced minima at ~20 K, and these minima, unlike those in "magnetic" (Kondo) alloys (see Chapter 6), show no broadening when the Ni content is increased. The second point of interest is that in the *ferromagnetic* (6% Ni) alloy the minimum at 20 K is completely absent. (The maxima at ~70 K in the more dilute alloys can be attributed to phonon drag, and the evidence from this interesting set of alloys seems to be rather against the ferromagnetic anomalies discussed above.) Very similar behavior has been observed in the Ir–Fe alloys except that in this case the low-temperature peaks (at ~30 K) are positive. Nagasawa measured a single alloy, Rh + 0.7 at.% Fe, and the thermopower was still increasing negatively as T decreased. His lower limit was ~1.7 K. Graebner et al. extended the measurements on this system into the millikelvin range. A T^2 variation in the resistivity occurs well below 0.1 K, and interaction effects are evident. The thermopower data exhibit minima near 3 K; Rh + 0.1 at.% Fe produces approximately -10 μV/K at 2.55 K. These data do not extend beyond 5 K, and thus a detailed comparison of these minima with those of Pd–Ni is not possible. However, one feature is clearly different. Figure 5.8 indicates the absolute magnitude of the minimum for Pd–Ni increases as the Ni concentration increases; the absolute magnitude of the minimum for Rh–Fe decreases as the Fe concentration increases. The only detailed analysis of the influence of spin fluctuations on S_d has been carried out for the localized fluctuations found in Al alloys containing magnetic impurities [74Z1] and is not directly applicable to the alloys discussed here.

5.2d   *Two-Current Conduction and Spin Mixing*

In a ferromagnetic metal at low temperatures there exist two independent parallel currents, one from either spin direction. Consequently, the impurity diffusion thermopower at low temperatures, S_{d0}^{LT}, is given by

$$S_{d0}^{LT} = \frac{S_{d0}(\uparrow) + rS_{d0}(\downarrow)}{1 + r} \qquad (5.12)$$

where $r = \sigma_0(\downarrow)/\sigma_0(\uparrow)$, the ratio of the two impurity conductivities. Thus, as in any other dilute alloy, the observed S_{d0}^{LT} depends on the type but not on the concentration of the alloying element present. However, at high temperatures the conductivities in the two spin directions equalize owing to spin mixing, and the high-temperature impurity-diffusion thermopower S_{d0}^{HT} is given by

$$S_{d0}^{HT} = \frac{1}{2}[S_{d0}(\uparrow) + S_{d0}(\downarrow)] \qquad (5.13)$$

Although $S_{d0}(\uparrow)$ and $S_{d0}(\downarrow)$ are both proportional to T, it is clear that S_{d0}^{LT}/T might be markedly different from S_{d0}^{HT}/T, giving rise to a possible extremal value of the S_{d0} vs. T curves at intermediate temperatures.

Furthermore, the impurity terms must be combined by an equation such as (5.10) with an ideal component of diffusion thermopower, S_d^i, and this too may be subject to spin mixing. It is clear when we consider these problems with those discussed in the preceding section that thermoelectric effects in ferromagnetic metals may be very complicated indeed.

Nevertheless, some progress can be made [70F2, 68L2]. If the Wiedemann–Franz law is valid, Eq. (5.10) may be rewritten as

$$S_d = \frac{\rho_0 S_{d0} + \rho_i S_d^i}{\rho_0 + \rho_i} \qquad (5.14)$$

where ρ_0 and ρ_i are, respectively, the impurity and ideal components of resistivity. Under conditions of spin mixing, ρ_0 varies with T and may be written at room temperature as $c[\rho_0^{LT}(1\%) + \Delta(1\%)]$, where c is the solute concentration, $\rho_0^{LT}(1\%)$ the residual (impurity) resistivity, and $\Delta(1\%)$ a deviation parameter, each evaluated for a 1% alloy. In that case Eq. (5.14) may be written

$$S_d = \frac{c[\rho_0^{LT}(1\%) + \Delta(1\%)]S_{do} + \rho_i S_d^i}{c[\rho_0^{LT}(1\%) + \Delta(1\%)] + \rho_i} \qquad (5.15)$$

On differentiating (5.15) and letting c tend to zero, we find

$$\left(\frac{dS_d}{dc}\right)_{c=0} = \frac{[\rho_0^{LT}(1\%) + \Delta(1\%)](S_{d0} - S_d^i)}{\rho_i} \qquad (5.16)$$

At room temperature, where spin mixing is reasonably effective, S_{d0} is given by (5.13), while the value of S_d^i is known from measurements on pure nickel. At helium temperatures, $\rho_i \ll \rho_0$ and $S_d \sim S_{d0}^{LT}$. Since $\rho_0^{LT}(1\%)$ and $\Delta(1\%)$ are both known from resistivity measurements, and since r can be obtained from (5.4) and (5.5) together with the assumption that $\Delta = \rho_0^{HT} - \rho_0^{LT}$, it is clear that (5.12) and (5.16) become simultaneous equations for $S_{d0}(\uparrow)$ and $S_{d0}(\downarrow)$. Analyses of this kind have been presented by Leonard [68L2] and Farrell and Greig [70F2] for nickel alloyed with $3d$ transition elements, and by Cadeville and Roussel [71C2] for nickel alloyed with $3d$ transition elements and for a wide range of cobalt alloys.

An alternative method for deriving $S_{d0}(\uparrow)$ and $S_{d0}(\downarrow)$ has been demonstrated by Cadeville et al. [69C3]. Instead of varying the temperature and introducing the ideal transport parameters ρ_i and S_d^i, these authors have measured S_{d0}^{LT} in ternary Ni–Co–Cr alloys and have used the relative concentrations of the two alloying elements as a variable coefficient in (5.12) and (5.14). In this way they are able to derive $S_{d0}(\uparrow)$ and $S_{d0}(\downarrow)$ for both Ni–Co and Ni–Cr.

Although the details of these various calculations differ, the general pattern that emerges is totally consistent. *Experimentally*, S_{d0}^{LT} changes sign from negative to positive as the valence difference between the solute and solvent is increased. An example of this for nickel alloyed with dilute concentrations of other $3d$ elements is shown in Figure 5.9. *Analytically* this is shown to result from a change in sign of $S_{d0}(\uparrow)$—the impurity thermopower in this majority spin direction—which can be attributed to a virtual bound state (see Chapter 6) passing through the Fermi level when the solute atom originates from the approximate midrange of the particular transition series.

As a final comment in this section, we should add that for many dilute ferromagnetic alloys, particularly those formed between neighboring

---

* In using this formula great care should be taken that any published values of dS/dc represent a true slope at $c = 0$, and not simply the change in S between the pure metal and a 1% alloy. Because of ambiguity over this point in earlier references, the values of $S_{d0}(\uparrow)$ and $S_{d0}(\downarrow)$ for Ni–Cr, Ni–V, and Ni–Ti quoted in [70F2] are wrong. For Ni–Co, Ni–Fe, and Ni–Cu, on the other hand, S is not too different from its value in pure nickel, and the difference in the definitions of dS/dc is unimportant.

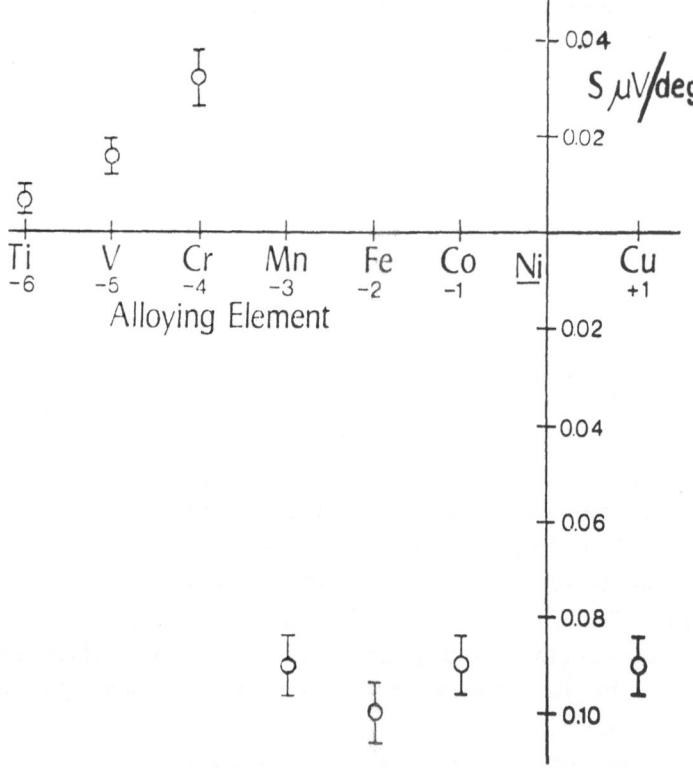

Fig. 5.9. Impurity thermopower of nickel alloyed with dilute concentrations
of other elements. Data are for 1 at.% concentrations ($T = 1$ K).

elements, $S_{d0}(\uparrow)$ and $S_{d0}(\downarrow)$ are roughly comparable. It therefore seems most
unlikely that spin-mixing effects could give rise to any recognizable extremal
values, so when these occur they must originate in some other mechanism.
Armstrong and Fletcher [72A2] also reach this conclusion in a study of
concentrated Ni–Fe alloys in which there are anomalies in the temperature
dependences of both ρ and S.

5.2e   *Curie Point Anomalies*

The anomalous behavior of the temperature dependence of the thermopower of nickel in the neighborhood of the Curie temperature T_c was
first observed by Tait in 1872. His results showed—and later measurements
have amply confirmed—that above room temperature the measured S in

nickel decreases linearly with T until it reaches a minimum (negative) value at about 250°C. Above that temperature, S rises with T up to $T \sim T_c$ (358°C) with a total increase of about 25%, and thereafter again decreases linearly with T as at lower temperatures. These general features of the measurements can be seen in Figures 5.3 and 5.5.

To explain this variation, Mott [36M1] proposed that the change in slope was a simple consequence of the equalization in energy with increasing T of the two sub-d-bands of opposite sign. He adapted his s–d scattering model to allow for two currents of opposite spin with $\tau(\uparrow) \propto 1/N(\uparrow)$ and $\tau(\downarrow) \propto 1/N(\downarrow)$, where $N(\uparrow)$ and $N(\downarrow)$ represent the effective densities of states at the Fermi energy in the two directions of spin. Within this approximation (5.9) becomes

$$S_d = -\frac{\pi^2 k^2 T}{3|e|}\left(\frac{3}{2\eta} - \frac{dN(\uparrow)/d\varepsilon + dN(\downarrow)/d\varepsilon}{N(\uparrow)+N(\downarrow)}\right)_{\varepsilon=\eta} \tag{5.18}$$

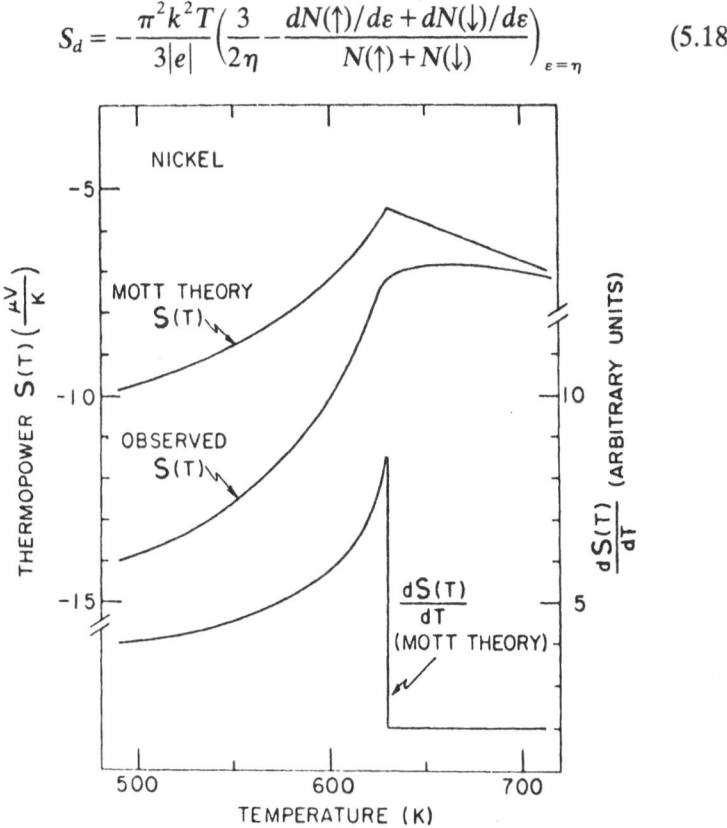

Fig. 5.10. The thermopower anomaly in nickel as predicted by Mott (Eq. 5.19) on the basis of s–d interband scattering (from [74T3]). Here, the "Observed $S(T)$" is the thermopower of a nickel–platinum *thermocouple* as measured by Grew [32G1], *not* the absolute TEP of pure nickel. See Fig. 3.10b.

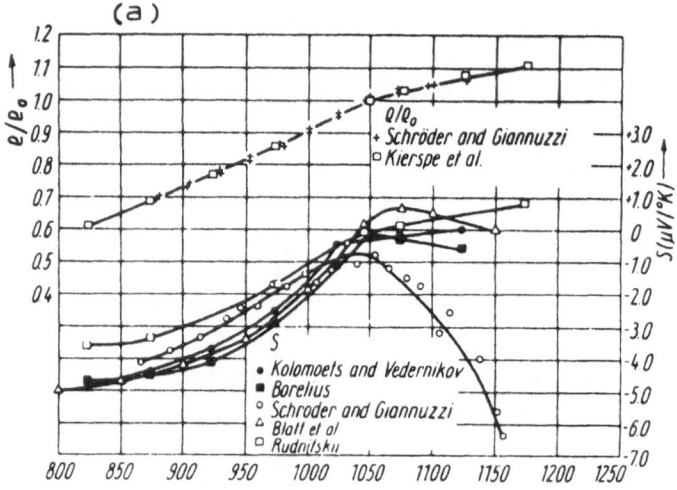

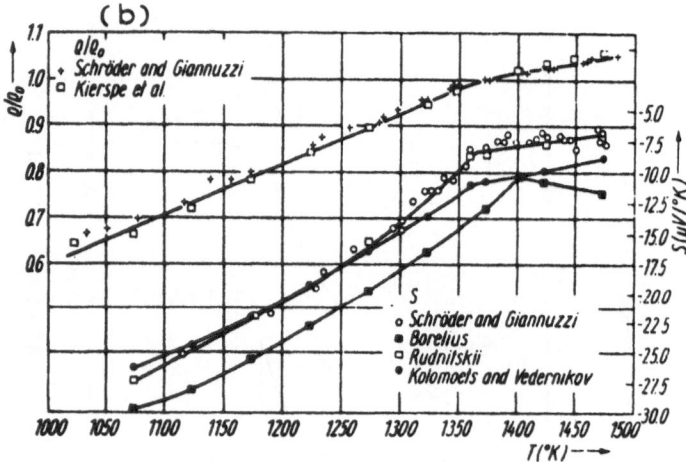

Fig. 5.11. Resistivity and thermoelectric power of (a) iron, (b) cobalt, and (c) nickel in the vicinity of their Curie temperatures (from [69S2]).

The changing occupation of the two sub-d-bands with T can be expressed in terms of $M_s(T)/M_s(0) = z$, the ratio of the saturation magnetization at temperature T to that at the absolute zero. For a band model we can say that at any T the fraction $\frac{1}{2}(1-z)$ of the unoccupied d-states has spin-↑, while the remainder, $\frac{1}{2}(1+z)$, has spin-↓. For parabolic d-bands it is then straightforward to express $N(↑)$ and $N(↓)$ in terms of z. The expression

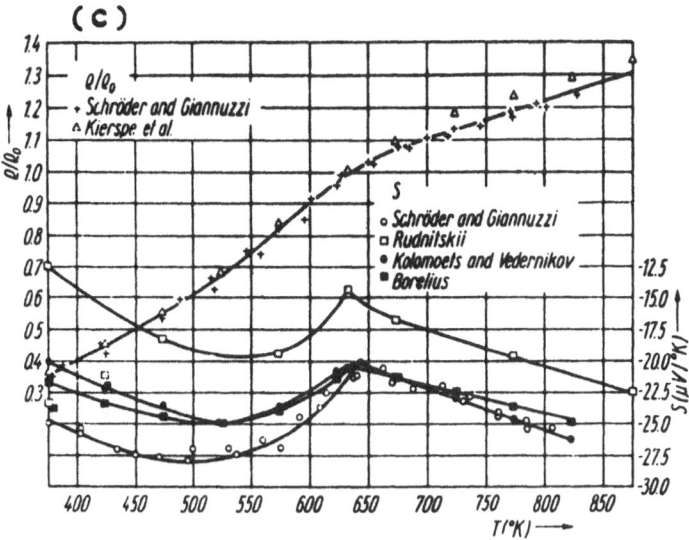

Fig. 5.11. *Continued.*

for S_d obtained by Mott on this basis is

$$S_d = -\text{const } T \frac{\left[\dfrac{1}{(1-z)^{1/3}}\dfrac{\alpha+(1+z)^{1/3}}{\alpha+(1-z)^{1/3}}+\dfrac{1}{(1+z)^{1/3}}\dfrac{\alpha+(1-z)^{1/3}}{\alpha+(1+z)^{1/3}}\right]}{2\alpha+(1+z)^{1/3}+(1-z)^{1/3}}$$

(5.19)

where $\alpha = \tau_{sd}/\tau_{ss}$. In Figure 5.10 we show the comparison of the experimental variation of S with T with that given by (5.19), and it is clear that the two curves are in reasonable qualitative agreement.

More recently Kolomoets and Vedernikov [62K1] have attempted to apply Eq. (5.18) to measurements of S above and below T_c in the case of all three ferromagnetic elements, iron and cobalt as well as nickel (see Figure 5.11). However, as the sub-d-bands of iron and cobalt contain fewer electrons than those of nickel, the simple parabolic band model is less appropriate, and the authors do not attempt a detailed comparison of theory and experiment. Nevertheless, they show that the overall features of the experimental curves—and these are quite complicated for iron and cobalt— can be explained qualitatively by the known band structure of these two metals.

The broad outline of the variation of S with T over a temperature range of several hundred degrees on both sides of T_c has also been investigated, for iron, cobalt, and nickel, by Schröder and Giannuzzi [69S2]. In their paper they compare their own measurements for all three elements with the best data generally available, and we reproduce these data in Figure 5.11. It is clear that the results are generally in good agreement with those of Kolomoets and Vedernikov [62K1], except that, for iron, Schröder and Giannuzzi find a more rapid decrease in S above T_c.

These authors have also estimated the change in Fermi energy $\Delta\eta$ due to ferromagnetic ordering from the corresponding changes in thermopower and resistivity, ΔS and $\Delta\rho$. The relationship, which follows from an expansion of the basic thermoelectric equation, is $\Delta\eta \approx (\pi^2 k^2 T/3e)(\Delta\rho/\rho)^2 \Delta S^{-1}$ [68S1]. Their values of $\Delta\eta$ are $0.157\ e$V, $0.053\ e$V, and $0.072\ e$V for Fe, Co, and Ni, respectively, although the Fe value in particular is very sensitive to the exact temperature at which the paramagnetic parameters are evaluated.

As we have mentioned previously in Section 5.1f, much of the recent work on thermoelectricity in the neighborhood of the Curie point has been concerned with the divergent behavior at temperatures within ± 20 K of T_{cj} that is, the region in which the temperature derivative of S can be expressed in the form of (5.6). As we see in Figure 5.10, the Mott model, while giving the overall features of the variation of S with T, is inadequate in detail, since for $T > T_c$ it would simply lead to $dS/dT = $ const. An improved picture is obtained by considering the short-range spin–spin fluctuations, although even then the exact physical origin of the thermopower anomalies has not been finally settled. Stated broadly, there are two rival models with the divergent behavior attributed to scattering at localized spin fluctuations on the one hand, and on the other to a reflection of the behavior of the specific heat—the "static entropy" model.

The relationship between entropy—and hence specific heat—and thermopower has been discussed in many textbooks, and indeed Lord Kelvin (William Thomson) referred to the Thomson heat as the "specific heat of electricity." Very crudely, if an electron in a metal is moved sufficiently slowly through a temperature gradient dT, the heat evolved dQ is given by $-\mu e dT$ [cf. Eq. (1.7)], where μ is the Thomson coefficient that is related to S through the Kelvin relation $\mu = TdS/dT$ [cf. Eq. (1.8a)]. Moreover, since the change in entropy $d\mathfrak{S}$ associated with dQ is given by dQ/T, it is clear that the thermopower S can be identified with this "transport entropy" $\mathfrak{S}$, i.e., $eS = -\mathfrak{S}$. Consequently, the specific heat C_e of the electron must be related to S by

$$C_e = T\frac{d\mathfrak{S}}{dT} = -eT\frac{dS}{dT} \qquad (5.20)$$

which, for a metal in which S is proportional to T, is simply

$$S = -\frac{C_e}{e} \simeq -\left(\frac{k}{e}\right)\left(\frac{kT}{\eta}\right) \tag{5.21}$$

Unfortunately such an elementary analysis cannot be used to explain the variety of magnitudes of S found in different metals; for a number of common metals, S even has the wrong sign. The reason is that no account has been taken of scattering effects whereby hot electrons may be scattered quite differently compared to cold ones; that is, in (5.20) and (5.21) the transport aspects of thermoelectricity are neglected (see Chapter 2). Whether or not the measured values of S can be explained by (5.20) therefore depends on how closely the transport entropy can be identified with the static entropy.

This general approach once enjoyed considerable support and was first applied to nickel by Grew [32G1], who showed that between T_c and $0.7 T_c$ the measured specific heat followed that obtained from (5.20) tolerably well. However, after 1936 when Mott introduced his $s-d$ scattering model, these ideas lost favor until very recently, when they were resurrected by Tang *et al.* [71T1]. They showed that from 20 K below to 20 K above the Curie point of pure nickel, the value of specific heat obtained from (5.20) was in remarkably good agreement with the measured magnetic contribution to C_e. The two sets of data are shown in Figure 5.12, and the authors emphasize that in obtaining this diagram, no adjustable parameters were used. They conclude that the excellent correlation for this system arises because it is an "extreme itinerant-electron ferromagnet." At first sight this may seem rather surprising, particularly in view of the models discussed in earlier sections, whereby the itinerant electrons are s-like while the high specific heat in transition metals is characteristically d-like. However, in a later paper Tang *et al.* [74T3] discussed this point in detail and concluded that close to T_c both the specific heat and the temperature derivative of thermopower are determined by the same spin–spin correlation function, and that it is the static properties of the critical fluctuations that govern these anomalies.

A somewhat different picture is given by Parks and his colleagues [72T1, 73Z1], who assume that the critical thermopower arises from Born *scattering* of the conduction electrons by localized critical fluctuations; that is, they emphasize the transport aspect of thermoelectricity. However, since the critical behavior is again characterized by the same spin–spin correlation function, and since Fisher and Langer [68F2] had shown earlier by a similar argument that the critical component of $d\rho/dT$ should vary with T in the

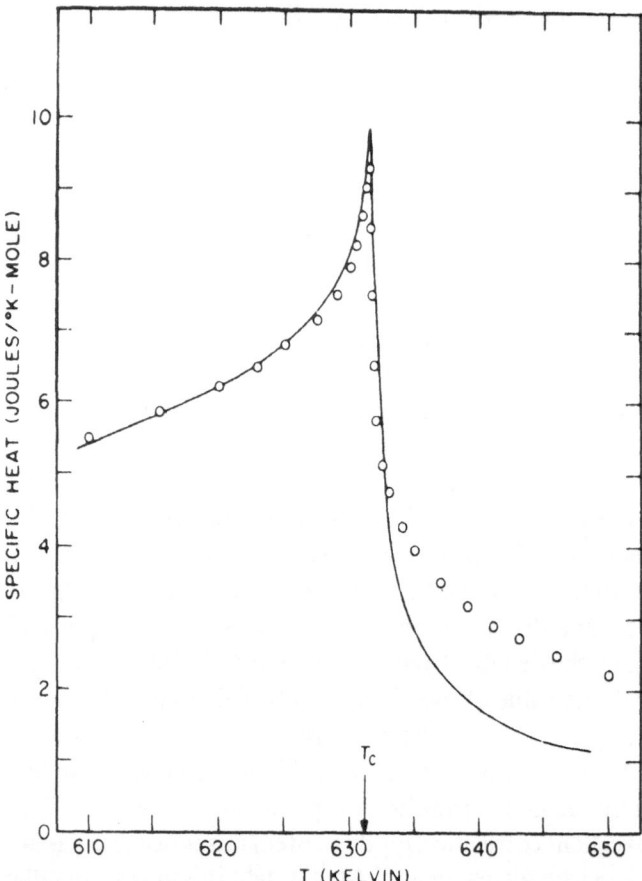

Fig. 5.12. Comparison of the magnetic contribution to the specific heat of nickel near the Curie temperature as deduced from thermopower measurements (solid curve) and conventional specific heat measurements (individual points) (from [71T1]).

same way as the specific heat, it is clear that a definitive separation of the two models is not easy. As evidence in favor of the scattering theory, Zoric *et al.* [73Z1] point out that in the ferromagnetic compound GdNi$_2$, dS/dT passes through a minimum a few degrees above T_c, and this is not reflected in the behavior of the specific heat. On the other hand, Tang *et al.* [74T3] argue that the *small* magnitude of the anomaly in dS/dT at the critical temperature of the order–disorder system, β-brass, is a point in favor of the static entropy model. The reason is that the change in entropy of that system is not electronic in origin but arises through a reordering of the ions, so that on their model the anomaly in dS/dT should be much smaller than C_e as, indeed, is observed.

5.3   The Phonon-Drag Thermopower of Transition Metals

As in "simple" metals, the S vs. T curves in transition metals all exhibit, at low temperatures, extremal values that are normally attributed to phonon-drag peaks. Again, as in nontransition metals, the peaks are positive in some metals and negative in others and may have the same sign as the diffusion thermopower (e.g., V, Ni) or the opposite sign (e.g., Mo, W, Pd, and Pt). We have already discussed in Section 5.2c how, for ferromagnetic metals, these peaks could be attributed to an entirely different mechanism, and in the next section we shall raise the question of yet another possible process that could give rise to peaks at low temperatures, namely, magnon drag. However, for the present it is worth emphasizing that these peaks are observed in all pure transition metals, so that any explanation that depends entirely on the existence of ferromagnetism is most unlikely. We shall therefore assume that these extremal values between 20 and 100 K are indeed attributable to phonon drag.

It should also be made clear that in transition metals there has been very little effort to study this behavior in detail. No attempts have been made to carry out careful measurements around 4 K and below of the type outlined in Chapter 4; that is, measurements made specifically to test the relationship $S = aT + bT^3$. Nevertheless, as a general trend we can say that the size of the maxima in transition metals is approximately an order of magnitude greater than those in simple metals, so that the magnitude of the phonon-drag thermopower, S_g, seems to scale with the values of S_d. Examples of the magnitudes of S_g have been given by Blood and Greig [72B3], who showed that, for a range of alloys based on Pd and Pt, the values of b obtained by plotting S/T against T^2 from about 1 to 15 K are between 2 and 8×10^{-10} V deg^{-4}; that is, roughly an order of magnitude greater than the values found by Guénault [74G5] in dilute copper alloys. Similarly, for pure nickel, $b \sim -3 \times 10^{-10}$ V deg^{-4} [65G1]. It is interesting to note that for Pd and Pt the values of $C_g/3n_0e$, the simplest theoretical estimate of S_g, Eq. (2.35), are also about an order of magnitude greater than the values in the noble metals. This may be because of the relatively small values of the density of s-electrons in these group VIII transition metals. We infer that this is the reason for the high values of S_g in these metals, but unfortunately there are insufficient data available to ascertain whether or not this is a general rule in transition metals.

With regard to the sign of S_g, Greig and his collaborators have shown in a series of papers that for a wide range of Pd and Pt alloys it is always positive [68F3, 72B3, 72G1]. This fact they have attributed to phonon-induced s–d or d–d electron transitions. It is rather difficult to choose between these alternative mechanisms, although the evidence is slightly in favor of the latter. The reason is that the greatest value of S_g is seen in Pt–Pd, which, of all alloys based on Pd or Pt, is that in which the hole current—the d-current—is likely to make the greatest contribution to the conductivity. Further evidence for this comes from the work of Fletcher *et al.* [70F3], who found that in β-phase Pd–H, S_g is negative. For such alloys the d-band is full, and the suggestion is that the negative S_g is the result of phonon-induced s–s processes.

Another feature of the phonon-drag peaks in Pd and Pt alloys is that the rate of quenching of S_g on alloying does not appear to depend on the mass difference between the solvent and solute atoms. For example, in Pd–Ru and Pd–Ir—two alloy series in which the atomic masses M of the solute atoms differ by a factor of almost 2, and in which the $(\Delta M/M)^2$ differ by a factor of almost 300—the reduction of S_g with increasing solute concentration is practically identical [72G1]. This point has been discussed in detail by Fletcher and Greig [68F3], who conclude that for these materials scattering at the strain fields around impurities seems to be a more important mechanism for reducing the phonon current than scattering due to the difference in the atomic masses (see Section 4.6).

Carter *et al.* [70C1] have investigated the low-temperature thermopower of the eight metals V, Nb, Ta, Mo, W, Rh, Ir, and Re and have commented briefly on the sign of S_g in each. In the group V metals (V, Nb, and Ta) and in Re, S_g is positive. This they attribute quite simply to phonon scattering by holes, since at low temperatures only the relatively small hole pockets can be spanned by the dominant phonons. In Mo and W the situation is rather more complicated. The main peaks at $\sim\theta_D/5$ are negative in both cases, but in W there is also a positive peak at ~10 K. Carter *et al.* interpret these observations as showing a rather delicate balance of electron and hole scattering across the various sheets of the Fermi surfaces and conclude that this balance changes with increasing T. In both Rh and Ir a negative peak appears at ~30 K followed by a positive peak at ~120 K. But Sellmyer and Franz [74S1], who measured S of Ir, observed only a positive peak near 140 K, which is, however, about three times as great as reported

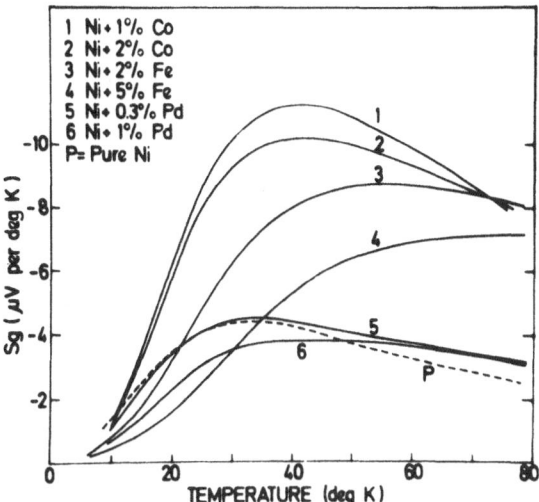

Fig. 5.13. Temperature variation of the phonon-drag thermopower in pure nickel and various dilute nickel alloys (from [68F5]).

by Carter *et al.* Similarly, Huntley [71H1] finds no indication of a negative peak in Rh. Therefore these results suggest that the samples (nominally 99.95%) of Carter *et al.* may have contained impurities with negative intrinsic thermopowers.

With reference to the phonon drag in ferromagnetic metals, most detailed experimental studies have been made on Ni and its alloys [65G1, 68F5, 70F2]. The most interesting feature of the results is that for a number of different nickel-based alloys the phonon-drag peak is appreciably *greater* than in the pure metal. In Figure 5.13 the values of S_g are shown in a number of different alloys—values obtained by subtracting the S_d estimated by a linear interpolation of low- and high-temperature measurements from the measured S—and for Ni–Co and Ni–Fe the enhancement is clearly observed. On the other hand, we see that for Ni–Pd, S_g is not much different from that in the pure metal.

The explanation of this effect is centered on the adaptation of Eq. (4.37) to the case of two-current ferromagnetic metals. For these the overall thermopower is given by

$$S_g = \frac{\sigma(\uparrow)S_g(\uparrow) + \sigma(\downarrow)S_g(\downarrow)}{\sigma(\uparrow) + \sigma(\downarrow)} \qquad (5.22)$$

where $S_g(\uparrow)$ and $S_g(\downarrow)$ are the characteristic phonon-drag thermopowers associated with the two spin directions. If it is assumed that $S_g(\uparrow)$ and $S_g(\downarrow)$ are neither subject to spin mixing nor substantially changed on alloying, then it is clear that the value of S_g is critically dependent on the ratio $\sigma(\uparrow)/\sigma(\downarrow)$. In Sections 5.1c and 5.2d we have pointed out how, for resistivities determined by impurity scattering, this ratio can change substantially from one alloy to the next. For Ni–Co, Ni–Fe, and Ni–Pd, the alloys shown in Figure 5.13, the ratios are of order 20, 15, and 1, respectively [68F1, 71F1], and these are in the same order as the magnitudes of the maxima in S_g. The modest reduction of $|S_g|$ on alloying in any one series is, of course, due to the usual decrease in phonon current in the presence of impurities.

It is clear that this provides an alternative explanation of the low-temperature maxima in nickel alloys to that given in Section 5.2c, where the emphasis was on anomalies in the diffusion thermopower S_d. At present, there seems no way to choose between these two proposals, and we merely restate the point made earlier in the section, namely, that it would be surprising if there were no phonon-drag thermopower in dilute nickel alloys and if any anomalous S_d appeared in virtually the same temperature range as the phonon-drag maxima in pure nickel.

Although systematic measurements of this sort have not been attempted for iron, an interesting effect of quite a different nature is observed. In pure iron, and in many dilute iron alloys, a well-defined positive maximum exists in S at ~ 200 K [67B4], which is roughly equal to $\theta_D/2$ for that metal. Such a temperature is rather too high to be associated with phonon drag, which characteristically appears at $\sim \theta_D/5$, and furthermore the peak is not greatly reduced either by cold-working or by alloying. Blatt et al. [67B4] initially attributed these peaks to the effect of magnon drag, and we shall again refer to this in the next section. However, in making a comparison recently of all possible sources of this peak, Blatt [72B5] concluded that phonon drag ought not to be neglected. In transition metals electron–phonon interactions are stronger than in simple metals, so that S_g can continue to increase to comparatively high temperatures. Evidence of this effect is clearly seen in certain Pd and Pt alloys [68F3].

5.4 Magnon Drag

In ferromagnetic metals and alloys the possibility exists that electrons may be driven along the temperature gradient by the magnon thermal current, thereby creating an additional thermopower, known as *magnon drag*, S_m. The effect of magnon drag, which is exactly analogous to the better-known phenomenon of phonon drag, was first discussed by Bailyn [62B1]. He concluded that the analytical expression for S_m is very similar to that for S_g, and that distinguishing the two drag effects could be extremely difficult. However, as a first approximation to the magnitude of S_m, we can write by analogy with equation (2.35)

$$S_m \simeq C_m/3n_0e \qquad (5.23)$$

where C_m is the magnon specific heat. As C_m varies as $T^{3/2}$ we see that, in principle, S_m can be distinguished from S_g (which to the same approximation varies as T^3), provided the magnitude of the effect is sufficiently great [64G1, 64R1]. Unfortunately, all estimates of S_m lead to the conclusion that it is extremely small. For example, for both Ni and Fe at 50 K, the magnon specific heat is only ~ 10 mJ mole^{-1} deg^{-1} in comparison to a lattice specific heat ~ 2.4 J mole^{-1} deg^{-1} [65D1]. On this elementary basis it is clear that S_m is likely to be more than 200 times smaller than phonon-drag effects at temperatures at which the latter are most prominent. Likewise Bhandari and Verma [69B1] have argued in a paper in which they gave a detailed derivation of (5.23) that S_m for Gd is likely to be only 5% of S_g and, therefore, quite negligible. They emphasize, however, that in (5.23), Umklapp processes are neglected.

In spite of these predictions much effort has gone into the search for magnon-drag effects in ferromagnetic metals and alloys. As mentioned in the preceding section, Blatt *et al.* [67B4] attributed the large peak in Fe at 200 K to magnon drag. In putting forward this suggestion, they discounted phonon-drag effects because the magnitude of the peak is neither reduced by cold-working nor greatly affected by alloying. Furthermore, they observed a $T^{3/2}$ component in S between 15 and 70 K. However, MacInnes and Schröder [71M2] have recently shown that the maximum can also be accounted for by a two-band model in the presence of a large magnetic field—this field simulating the effect of spin–orbit interactions on the conduction electrons. Even more recently Blatt [72B5] measured the field dependence of thermopower in pure Fe and found that the changes observed are both too small and of the wrong sign to be accounted for by MacInnes and Schröder's model. Blatt concluded that the thermoelectric behavior of iron is still unexplained and listed spin–orbit scattering, electron–electron scattering, phonon drag, and magnon drag as possible origins of the peak at 200 K.

However, magnon drag appears to be of some significance at low temperatures (5 to 25 K) in antiferromagnetic chromium. Unfortunately, the dispersion relation for magnons in this antiferromagnet is the same as for phonons leading to the same T^3 dependence for both phonon and magnon drag. Hence, it is not possible to separate the two contributions by a careful study of the temperature dependence of the thermopower. Based on the behavior of the thermopower in dilute Cr–Mn and Cr–V alloys, Trego and Mackintosh [68T2] conclude that magnon drag makes a substantial contribution in the Cr–Mn system.

A more recent claim to the identification of magnon drag was made by Granneman and Berger [72G2] from their measurements of thermoelectric effects in *concentrated* nickel alloys. They measured the magnetic-field

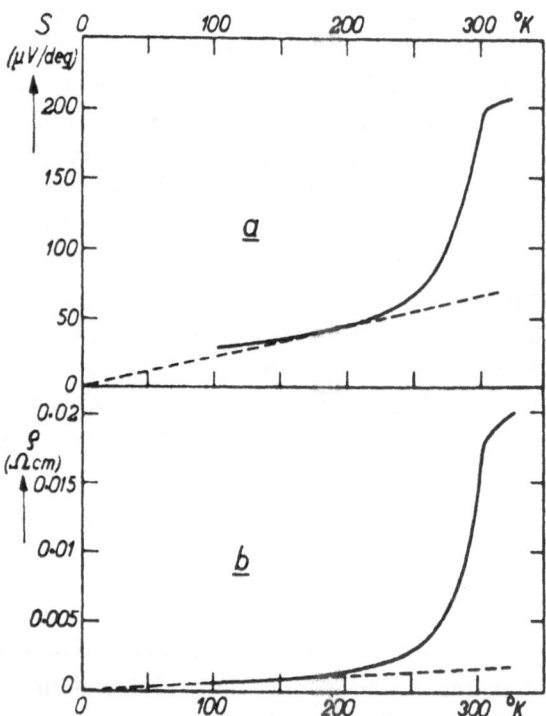

Fig. 5.14. Effect of magnon–electron scattering in degenerate p-type MnTe on (a) thermopower and (b) resistivity. The dashed lines represent nonmagnetic contributions $S = 0.23T\,\mu V/\text{deg}$ and $\rho = 6\times10^{-6}T\,\Omega\text{-cm}$, respectively (from [64W1]).

dependence of the Peltier coefficient Π in $Ni_{70}Cu_{30}$ and $Ni_{70}Fe_{30}$ between 1.3 and 4.2 K, and found that, on increasing the applied field to 5.5 T, the observed Π fell by ~5%. They ascribed this to the magnon-drag effect's being quenched by the field. The advantage of using such concentrated alloys is that the phonon-drag thermopower can probably be neglected.

Finally, we should add that the only other instance in which an anomalous thermoelectric effect has been ascribed to magnon drag is in the case of an antiferromagnetic semiconductor, MnTe [64W1]. In that compound the measured S rises rapidly between 200 K and the Néel temperature 310 K, increasing from $\sim 50\,\mu V\;deg^{-1}$ to $\sim 200\,\mu V\;deg^{-1}$ (Figure 5.14). The magnitude of the effect is due partly to the strong interaction between the current carriers and the ion spins, and partly to the fact that in semiconductors the number of carriers—the n_0 in the denominator in (5.23)—is much less than in metals.* For a detailed account of the influence of magnon drag on the transport properties of magnetic semiconductors, see the recent review by Haas [73H1].

5.5 Transition Elements: Summary of Experimental Results

There are 24 transition metals originating from 8 groups in the periodic table. Elements within each group have many common physical and chemical properties, and it has been emphasized by Vedernikov [69V1] that nowhere is this more true than for thermoelectric effects at high temperatures (see Figure 5.5). His review article ends with the following statement,

The temperature dependence of the thermoelectric power of transition metals in a given (any) group of the periodic system, if the metals are in a magneto-disordered state, may be expressed in the range above 80 K by a single $S(T)$ curve, typical for the group, whose shape does not depend on the crystal structure.

Vedernikov's article was devoted entirely to high-temperature aspects of thermoelectricity, and he did not consider measurements below 80 K. However, from the data discussed in previous sections, it is clear that the same general statement applies, although at these temperatures complications of magnetic anomalies are more likely. For example, we have seen that whereas the phonon-drag peaks of Pd and Pt are positive, that of Ni is negative. Of course, at very low temperatures the data are characterized

---

* We recall that it is for this reason that S_g is also very large in semiconductors.

both in sign and magnitude by the residual impurity present, so that at helium temperatures the values of S/T are generally not characteristic of the parent transition metal.

Although there are these common features in the data from the three elements in each group, the measurements from the various groups have little in common. Some elements give values of S that are large and positive, whereas for others S is equally large and negative. We have discussed this in terms of the general trend across the periodic table in Section 5.2a, but it is clear from Figure 5.5 that there is a detailed variation from group to group which is rather difficult to summarize. However, as a broad statement we can say this: For most transition metals there is (a) an extremal value of the S vs. T curve between 50 and 100 K that we attribute to a phonon-drag peak. (b) There is a region in which S varies linearly with T, starting between 200 and 300 K and extending upward by sometimes as much as 1000 K. (c) There is a further extremal value of S at high temperatures. This second peak is normally thought to be due to the temperature modification of the mean density of available d-states, $N_d(\varepsilon)$, within a range kT of the Fermi energy (see Section 5.2a and reference [70A1]). However, there is the alternative possibility that it might be due to electron–electron effects [71C1]. (d) A fairly sharp break in the S vs. T curve, may occur when any one element undergoes a structural or magnetic transformation. From a review of the data we have attempted to summarize in Table 5.1 the presence or lack of these four features for each transition metal.

In Table 5.2 we have listed the low-temperature values of S at a number of selected temperatures for each pure transition metal. Many of the data are taken from Vedernikov's article [69V1], which is not always the best source, and a number of other references are included. For virtually all elements the measurements have been made on polycrystalline wires, with no attempt, except in the case of yttrium, to investigate anisotropy effects in the cph metals. For yttrium the results obtained by Sill and Legvold [65S2] on a single-crystal specimen seem significantly better than any others on that metal, and we have therefore included these in Table 5.2. A small anisotropy in S between the basal plane and the $\langle 0001 \rangle$ direction is apparent.

Many of the measurements made on the elements have been carried out solely to obtain data or, as in the case of yttrium, to illustrate some simple qualitative effect. Consequently, there is little that we can add by way of a general account that has not already been discussed fully in earlier sections. However, there are two quite separate sets of elements that have been studied more thoroughly than the others. These are (a) the "magnetic" metals and (b) the metals that are commonly used as the elements of thermocouples.

5.5a   The Magnetic Elements: Cr, Mn, Fe, Co, and Ni

There are five magnetic elements, all from the $3d$ transition series, three of which are ferromagnetic (Fe, Co, and Ni) and the remaining two are antiferromagnetic (Cr and Mn). Although we have already discussed in detail the temperature variation of S close to the Curie point in the ferromagnetic metals, there are also a number of general features of the thermopower of all five elements that are most usefully discussed together. Therefore, in this concluding section we shall make a broad survey of these features from a slightly more experimental point of view.

Iron is bcc both at normal temperatures and close to its melting point but is fcc in an intermediate temperature range between 1183 and 1663 K. The thermopower of iron has been thoroughly investigated, and Vedernikov [69V1] shows that there is general agreement about the overall form of the results. There is a maximum in $S \sim 18\ \mu\text{V deg}^{-1}$ at 200 K, followed by a minimum $\sim -5\ \mu\text{V deg}^{-1}$ at 800 K, and there is a very sharp change in the values of S at both structural transition points. It is the maximum at 200 K that has been variously ascribed to magnon drag, two-band conduction in the presence of a large internal field, and phonon drag (see Section 5.4). For iron, the change in dS/dT at T_c (1043 K) tends to be obscured by the sharp fall in S of $\sim 6\ \mu\text{V deg}^{-1}$ at the $\alpha-\beta$ transition (1183 K), although the detailed measurements shown in Figure 5.11 illustrate the change in dS/dT quite clearly.

For both Co and Ni there are no such complications. Both metals have well-established fcc structures at their Curie points—1393 K and 631 K, respectively—and neither undergoes further phase changes with increasing T. However, although the stable form of Co at room temperature is cph, it is found that at normal temperatures the cph and fcc forms coexist [58P1]. In general, the larger the grain size, the greater the proportion of hexagonal material in the metal. The structure becomes entirely fcc above about 723 K, at which point there is only a 10% change in the value of S. Furthermore, and rather surprisingly, there is no widespread variation in S at low temperatures resulting from possible differences in the proportions of cph and fcc material in the specimens used by different workers. For Co it appears therefore that the crystal structure is not an important parameter in determining S. It is of greater interest to note that the S vs. T curve goes

Table 5.1. Salient Features of the Thermopower of Transition Elements

Element	L.T.Max/Min value (μV deg^{-1})	Temp. (K)	Linear region	H.T.Max/Min value (μV deg^{-1})	Temp. (K)	Changes in S attributed to structural or magnetic transformations
Group IIIB						
Sc	−7.5	150	150–500 K and 750–1400 K	+10.5	1550	
Y	−5.0	90	300–900 K	+4.2	1400	
La	Not. meas.		300–583 K	−1.5	At M.P. (1193)	Change in slope at α(cph) – β(fcc) transition temperature.
Group IVB						
Ti	−3.0	80	400–800 K and 1500–MP(1941 K)	+7.0 and −3.0	400 and 950	Discontinuity at α(cph) – β(bcc) trans. temp. observed by one author [51W1] but not by another [69V1].
Zr	Insuff. data		400–800 K 900–1400 K and 1400–1800 K	+9.0 and −6.0	300 and 800	No change observed at α(cph) – β(bcc) transition.
Hf	Not. meas.		500–1000 K and 1100–1600 K	+8.0	500	No data available.

Group VB						
V	+3.0	75	500–1100 K	+1.0 and +9.5	400 and 1300	Small discontinuity at 217 K.
Nb	+3.1	75	600–1400 K	−1.5 and +6.0	500 and 1400	
Ta	+1.5	40	1100–1800 K			
Group VIB						
Cr	+9.6	40	500–800 K	+19	900	Pronounced maximum a few degrees below the Néel temperature.
Mo	−0.5	50	200–600 K and 1200–2100 K	+17	1100	
W	−4.5	50	200–800 K and 1800–2400 K	+19	1400	
Group VIIB						
Mn	+16	35	300–1000 K	+3	1100	Sharp increase in S (a) at and below Néel temp.; (b) at and above α(cubic) − β(cubic) transition temperature.
Tc	–	–	–	–	–	
Re	+2	30	450–1500 K	+7	1650	

continued

Table 5.1. *Continued*

Element	L.T.Max/Min value ($\mu V\,deg^{-1}$)	Temp. (K)	Linear region	H.T.Max/Min value ($\mu V\,deg^{-1}$)	Temp. (K)	Changes in S attributed to structural or magnetic transformations
Group VIIIa						
Fe	+17	200	300–700 K	−5	800	Small reduction in slope of SvT curve at Curie point. Sharp reduction in S at α(bcc) − γ(fcc) transition; sharp increase in S at γ(fcc) − δ(bcc) transition.
Ru	−1.5[a]	200	500–1800 K	None obsvd.		
Os	−4.0[a]	200	600–1800 K	None obsvd.		
Group VIIIb						
Co	−	−	100–500 K and 1400–1600 K	−48	600	Marked reduction of slope of SvT curve at Curie point. Small (10%) reduction in S at cph transition.
Rh	−0.15 and +0.8	30 and 130	600–1300 K and 1300–1700 K	None obsvd.		
Ir	−0.4 and +0.95	30 and 140	300–1300 K and 1400–1800 K	None obsvd.		

Group VIIIc					Maximum in S at Curie point.
Ni	-ve	35	200–500 K and 600–1500 K	None obsvd.	
Pd	+4.5	55	300–1800 K	None obsvd.	
Pd	+6.0	55	400–2000 K	None obsvd.	

[a] No measurements have been made below 80 K in these metals, so further extremal points could exist.

Table 5.2. Values of S in Pure Transition Metals below Room Temperature ($\mu V\,deg^{-1}$)

Element	Orientation	10 K	20 K	50 K	80 K	100 K	150 K	200 K	250 K	273 K	References
Group IIIB											
Sc	Ave	−1.6	−3.0	−8.0	−12.7	−14.0	−15.6	−16.3	−16.5	−16.2	64N1, 69M1, 69V1
	Basal	−0.3	−0.4	−3.3	−4.5	−4.4	−3.1	−1.9	−0.9	−0.7	59J1, 61B1, 65S2, 69T3,
Y	c-axis	−0.5	−0.4	−2.8	−4.5	−4.9	−4.6	−3.2	−1.4	−0.5	69V1
La	Ave	—	—	—	0	+0.3	+0.4	+0.7	+1.0	+1.3	61B1, 69V1
Group IVB											
Ti	Ave	—	—	−3.0	−3.0	−2.6	0	+2.0	+4.0	+4.5	41P1, 51W1, 69V1
Zr	Ave	—	—	0	+3.0	+4.5	+7.5	+8.5	+9.5	+9.5	41P1, 52A1, 69V1
Hf	Ave	—	—	—	0	0	+2.5	+3.7	+4.7	+5.3	69V1
Group VB											
V	Ave	+0.19	+0.76	+2.45	+2.91	+2.65	+1.52	+0.72	+0.26	+0.13	41P1, 63M1, 67S2, 70C1
Nb	Ave	+0.31	+0.98	+2.73	+3.09	+3.13	+1.42	+0.65	−0.04	−0.20	41P1, 65W1, 70C1
Ta	Ave	+0.36	+1.03	+1.41	+0.78	0	−0.8	−1.5	−2.0	−2.2	41P1, 70C1
Group VIB											
Cr	Ave	+3.1	+6.7	+8.2	+5.0	+5.0	+7.0	+11.8	+17.5	+18.8	41P1, 68S1, 68T2, 74G3
Mo	Ave	−0.02	−0.11	−0.48	−0.2	+0.1	+0.94	+2.50	+4.08	+4.57	41P1, 70C1
W	Ave	+0.05	−0.28	−2.78	−3.70	−4.04	−2.45	−1.41	−0.10	+0.56	41P1, 70C1, 73G2, 73T1, 74G2

										Ref.
Group VIIB										
Mn	Ave	—	+12.5	+15.5	+6.0	-2.5	-7.0	-8.5	-9.7	-10.0	63G3
Tc	Ave	—	—	—	—	—	—	—	—	—	—
Re	Ave	+0.61	+1.32	+1.18	+0.08	-0.66	-2.21	-3.51	-4.63	-5.03	69V1, 70C1, 72N1
Group IIIa										
Fe	Ave	+1.0	+2.5	+8.0	+12.0	+13.0	+16.0	+17.0	+15.5	+15.0	67B4, 70P1
Ru	Ave	—	—	—	+0.2	0	-1.1	-1.5	-1.5	-1.5	69V1
Os	Ave	—	—	—	-2.2	-3.2	-3.8	-4.0	-4.0	-4.0	69V1, 72N1
Group VIIIb										
Co	Ave	—	-0.5	-1.0	-3.0	-4.0	-9.0	-12.0	-18.0	-19.0	65R1, 69V1
Rh	Ave	-0.19	-0.33	-0.11	+0.54	+0.78	+0.92	+0.75	+0.58	+0.48	69V1, 70C1, 71H1
Ir	Ave	—	-0.11	+0.13	+0.57	+0.73	+0.77	+0.64	+0.46	+0.35	69V1, 70C1, 72N1, 74S1
Group VIIIc										
Ni	Ave	-2.0	-4.7	-7.2	-8.1	-8.4	-11.0	-13.5	-17.0	-18.0	65G1, 67B4, 69V1, 70F2
Pd	Ave	+0.4	+1.6	+4.3	+3.7	+2.00	-1.63	-4.85	-7.42	-9.00	58C2, 68F3
Pt	Ave	+0.6	+2.3	+5.8	+5.5	+4.29	+1.32	-1.27	-3.28	-4.45	58C2, 68F3

through a very deep minimum at 600 K and then abruptly changes slope at the Curie point. Rather few measurements have been made below room temperature, but in those that are available [see 69V1] one unusual feature is apparent. Unlike the situation in every other transition metal in which measurements are reasonably complete, there is no extremal value in S below room temperature.

The thermopower of Ni is known to a much better accuracy than that of the other "magnetic" metals, and, as summarized by Vedernikov, the agreement between the various sets of data is quite good. As discussed in earlier sections, the values of S are negative at all temperatures and decrease continuously with increasing T, except at temperatures just below T_c. In this region S rises with T and then begins to fall again as the temperature is increased still further (Figure 5.10).

Chromium is bcc at all temperatures, although there is some evidence that it transforms to fcc just below its melting point. The temperature dependence of S is fairly complicated for this metal, with three separate maxima at different parts of the temperature range. In the magnetically disordered state at high temperatures, there is reasonable agreement between the various sets of results [69V1, 67C3], with S fairly large and positive up to ~850 K, after which it drops slowly to near zero. Below the Néel point there are four sets of data [41P1, 68S1, 68T2, 74G6] down to ~100 K, showing a fairly sharp maximum in S just below T_N. At still lower temperatures the most reliable data appear to be those of Goff [74G6], who finds a minimum in S at 100 K and a maximum, presumably due to phonon drag, at 40 K. On annealing, this maximum increased in magnitude from 6.3 μV deg^{-1} to 9.6 μV deg^{-1}. Some of the earlier measurements on both ρ and S seemed to indicate that many of the data on chromium were markedly susceptible to hysteresis effects. However, recent workers no longer mention this problem, and it is worth emphasizing that the agreement between the four sets of measurements mentioned above, through a very pronounced peak just below T_N, is about the best agreement in thermopower measurements in any transition metal.

Of all such metals, manganese has by far the most complex crystal structure. The normal α-phase is cubic but with 58 atoms per unit cell. At higher temperatures, there are two transformations: to fcc at 1368 K and to bcc at 1406 K. In addition, there is the Néel point transition at 94 K. Consequently, Mn is quite unsuitable as a thermocouple element, and only two attempts have been made to measure its thermoelectric properties [69M1, 63G3], one at low temperatures and the other at high temperatures.

There is a rather sharp increase in S at the $\alpha-\beta$ transition (1000 K), but little evidence of discontinuities at the two higher transition points. Below room temperature there is a marked change in S from $-10\ \mu V\ deg^{-1}$ at 300 K to $15\ \mu V\ deg^{-1}$ at ~40 K. In particular, there is a rapid increase in S as the temperature decreases through T_N, exactly the same effect as was found in chromium. The electrical resistivity of Mn is by far the highest in any pure transition metal [65M1], and so it might be expected that the thermopower would also show quite unusual behavior. However, although there are certain anomalies present as mentioned above, the temperature dependence of S is not too unorthodox and is, in fact, rather like that of the other common VIIB metal, rhenium.

5.5b The "Thermocouple" Elements: Fe, Pt, Re, and W

To be useful as a thermocouple material, a metal or alloy must have at least six desirable properties. These are (1) a large value of thermal emf, E, which (2) is of opposite sign to that of its partner in the thermocouple circuit, and which (3) increases continuously with increasing T. In addition, the thermoelement should be (4) chemically stable, (5) easily fabricated and worked, and (6) reasonably inexpensive.

When all these factors are taken into consideration, together with the range of temperature over which measurements are required, it is found that four pure transition metals, Fe, Pt, Re, and W, are frequently used as thermoelements, while Ir, Mo, Pd, and Ta are used occasionally. The advantages and disadvantages of each metal are discussed in great detail in *Temperature, Its Measurement and Control in Science and Industry* [73I1], and only the briefest outline of these factors is given here. A summary of the main points is given in Table 5.3.

The oldest and most basic noble metal thermocouple is Pt vs. Pt + 10%Rh, the emf of which is used to define the International Temperature Scale between 630.74°C and 1064.43°C. The outstanding properties of the platinum metals and their alloys are high melting points, reproducibility, accuracy over a wide temperature range, resistance to corrosion, and stability in calibration. The disadvantages are that the emf generated in the thermocouple is not outstandingly high, making it of little use at low temperatures, and that the pure metal tends to become mechanically weak above 1400°C. Although it can be used in air or in an inert atmosphere, Pt cannot be used in a reducing atmosphere or in a vacuum. The problem in the

Table 5.3. Transition Elements Used in Thermocouples

Metal	Advantages	Disadvantages
Platinum	Excellent mechanical and chemical properties; high degree of homogeneity, and reproducibility. Stable in oxidizing atmosphere. Low electrical resistance.	Low thermal emf, especially at low temperatures. Unstable in vacuum. Unstable in reducing atmosphere.
Alternative to platinum		
Palladium	Higher thermal emf. Cheaper.	Lower melting point. Absorbs hydrogen. Surface oxides formed.
Iridium	Higher melting point.	Brittle in hydrogen. Surface oxides formed. Expensive.
Tungsten	Very high melting point. Stable in vacuum. Stable in hydrogen and inert gas atmospheres. Inexpensive.	Brittle wire, made worse by oxidizing. Decreasing thermopower above 1300 K.
Alternatives to tungsten		
Rhenium	Ductile.	Very expensive.
Tantalum	Ductile.	Readily absorbs residual gases.
Molybdenum	Inexpensive.	Lower thermal emf. Lower melting point. Forms nitrides as well as oxides.
Iron	Inexpensive. Ductile. Large emf with respect to constantan. Stable in oxidizing and reducing atmospheres.	Highest useable temperature ~750°C. Rusts.

latter case appears to be that the Pt absorbs the alloying element from the other limb of the thermocouple.

At any given T, the thermopower of Pd is roughly twice that of Pt [58C2], so that, as a sensitive noble metal thermoelement, Pd appears to be a better choice. On the other hand, Pd has a rather lower melting point than

Pt, possibly oxidizes in air above 700°C, and absorbs hydrogen in large quantities. Pure Ir has also been used as a thermoelement with the advantage that its melting point at 2410°C is more than 600°C above that of Pt. Unfortunately, surface oxides are formed on Ir at high temperatures, and the element becomes brittle in hydrogen. Furthermore, Ir is extremely expensive.

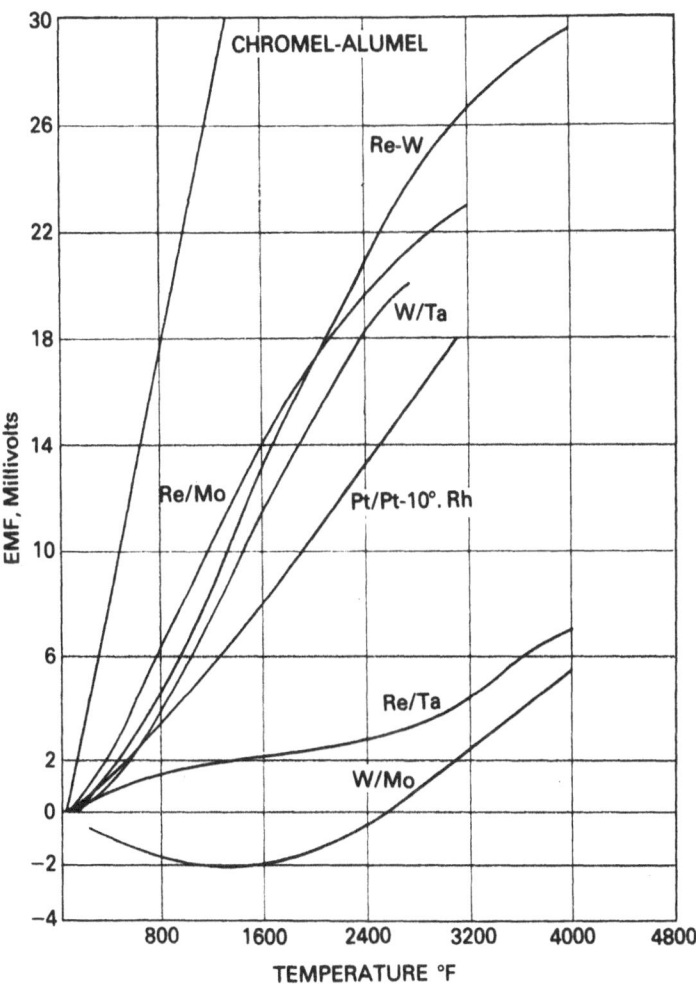

Fig. 5.15. Thermal emf of primary refractory metal thermocouples compared to Chromel–Alumel and platinum–platinum–rhodium types. The reference junction is at 0°C (from [62L1]).

For continuous thermometry above 1500°C, it is necessary to use refractory metals. The four pure metals most commonly used are W, Re, Ta, and Mo, and the temperature variation of E for thermocouples made from most combinations of these four metals is given in Figure 5.15. It is clear that the best characteristics are given by the W vs. Re combination, and this is a most useful thermocouple, even at lower temperatures in a vacuum and in hydrogen and inert gas atmospheres. Unfortunately, tungsten cannot be used in an oxidizing atmosphere; furthermore, the decreasing dE/dT above ~2500°F (1371°C) reduces the accuracy of the thermometer at the high-temperature end of the range. For this latter reason as well as the poor ductility of tungsten wire and the very high cost of Re, commercial thermocouples usually have *both* limbs made of W–Re alloys of different compositions.

As an alternative to W, Mo has little to offer. In addition to having a lower value of E at high temperatures (Figure 5.15), it has a lower melting point and, above 2400°C, it reacts with nitrogen to form nitrides. Ta has also been used as a thermoelement in combination with W but suffers from the disadvantage that it has a great affinity for gas molecules at high temperatures. Because of this, the value of E is variable and the thermoelement is unreliable. However, a W vs. Ta thermocouple has been successfully used in mercury vapor between 0 and 420°C [68F6].

In spite of the rather complicated temperature variation of its thermopower, the most widely used thermoelement of all is iron. The great advantages of iron are that it is readily available as inexpensive, homogeneous, ductile wires, it has a large value of E when used in conjunction with constantan, and it can be used under both oxidizing and reducing conditions. Again, although an iron vs. constantan thermocouple can only be used up to ~750°C, this is almost twice the range of a copper vs. constantan thermocouple in which the copper oxidizes very readily. Iron, of course, deteriorates very rapidly in a moist atmosphere.

5.6 Commercial Thermocouples

In this section we shall provide only a brief summary of a subject of great technological interest. For further details, the reader is referred to the recent book by Kinzie [73K1] and to *Temperature, Its Measurement and Control in Science and Industry* [73I1].

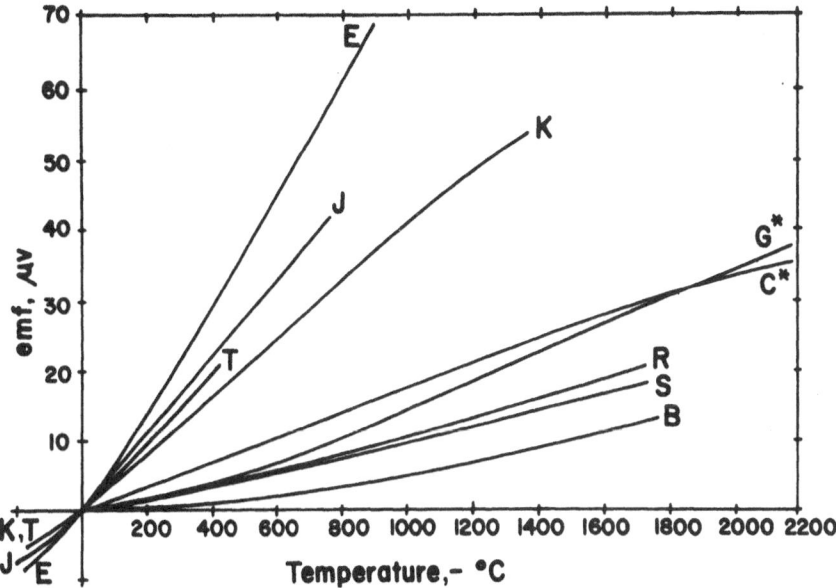

Fig. 5.16. The thermoelectric voltage of various common commercial thermocouples. The reference junction is at 0°C.

A practical thermocouple consists of two different metals or alloys with as large as possible a difference between their E vs. T characteristics. The only common combination of *pure* metals is W vs. Re. In general, one or both arms of the thermocouple is a transition metal *alloy*, chosen either for the large magnitude of E at a given T, or for its good chemical and metallurgical properties, or for some combination of these reasons. In addition to the six desirable qualities listed at the start of Section 5.5b, an alloy thermoelement must also be homogeneous and the value of $E(T)$ must not depend strongly on composition.

Although many combinations of metals and alloys have been investigated, there are only six common thermocouples in general use. These are copper vs. constantan (T), iron vs. constantan (J), chromel vs. constantan (E), chromel vs. alumel (K), tungsten vs. tungsten–rhenium (G* and C*), and platinum vs. platinum–rhodium (R, S, and B). The letters are the designations given by the American National Standards Institute and are in common use in the United States. Full details of the chemical and physical properties of all those thermocouples are given in references [73K1] and [73I1], so that in this account we shall only summarize the most characteristic features of each. The temperature variation of E for these six thermocouples with one junction held at 0°C is shown in Figure 5.16. Calibration tables are given in various NBS monographs [55S1, 72S1, 74P2] and in a recent publication by Omega Engineering, Inc. [71O1].

5.6a   Copper vs. Constantan (Type T)

Constantan is an alloy of copper and nickel of composition ranging from $Cu_{50}Ni_{50}$ to $Cu_{65}Ni_{35}$. The thermocouple grade is normally $Cu_{57}Ni_{43}$ with small quantities of impurities, such as Mn and Fe, added to ensure that the calibration remains in accordance with standard tables. Pure copper with low oxygen content is very homogeneous and, except at very low temperatures, has a highly reproducible thermoelectric power. The copper-constantan thermocouple is an inexpensive, accurate, and reliable means of determining temperatures from -260 to about 400°C, with the upper limit restricted by the oxidation of copper.

5.6b   Iron vs. Constantan (Type J)

As mentioned in the preceding section, iron vs. constantan is an extremely popular combination which can be used up to ~750°C; that is, almost twice the range of copper vs. constantan. Furthermore, it has the advantage that it can be used in both oxidizing and reducing conditions with slightly higher temperatures attainable in the latter case. However, since iron is less homogeneous than copper, this thermocouple has parasitic emfs along both limbs and is only roughly half as accurate as copper vs. constantan.

5.6c   Chromel vs. Constantan (Type E)

Chromel is an alloy of composition $Ni_{90}Cr_{10}$, which is sometimes specifically labeled chromel P. This particular thermocouple has the highest emf at a given temperature for any of the six major commercial thermocouples. As with iron vs. constantan, it can be used from ~-250°C to ~750°C and can be used in a mildly oxidizing or reducing atmosphere. The high thermoelectric power makes it extremely useful for differential temperature measurements.

5.6d   Chromel vs. Alumel (Type K)

Alumel is a rather complicated nickel alloy of composition Ni_{94}-Mn_3Al_2Si, developed as a thermoelement with a thermal emf of opposite sign to that of chromel. Although nickel itself could also be used for this purpose, the magnetic transition around 350°C ($T_c = 631$ K) makes it rather less satisfactory than the alloy. The chromel–alumel combination has a moderately large and reasonably constant thermoelectric power ~40 μV deg^{-1} between ~250 and ~1000°C, and has the advantage over the T, J, and E thermocouples in that it can be used up to 1300°C. It is also more resistant to oxidation than these other combinations and can be used up to ~1200°C without rapid deterioration. On the other hand, it cannot be placed in a reducing atmosphere. At low temperatures it has roughly the same characteristics as copper vs. constantan, although for measurements below room temperature another variety of chromel, chromel X ($Ni_{64}Fe_{25}Cr_{11}$), is recommended.

5.6e Tungsten vs. Tungsten–Rhenium (Types G* and C*)

As mentioned in the preceding section, these refractory metal thermocouples are used for measurements well above 2000°C. The pure metal combination tungsten vs. rhenium, although satisfactory up to ~1650°C, becomes less sensitive above that temperature, and the thermocouple is not very useful above ~2200°C. Furthermore, rhenium is very expensive. A better arrangement is to use tungsten–rhenium alloys, which at certain concentrations have higher thermal emfs than pure rhenium, are much more ductile, and are considerably cheaper. The most common thermocouples are tungsten vs. tungsten + 26% rhenium (type G*) and tungsten + 5% rhenium vs. tungsten + 26% rhenium (type C*). Both of these combinations can be used up to ~2700°C, with the latter more sensitive below 1800°C, and the former slightly more sensitive above that temperature. These thermocouples have poor oxidation resistance and must be used in a vacuum or in hydrogen or inert gas atmospheres.

5.6f Platinum vs. Platinum–Rhodium (Types R, S, and B)

There are three common types of these basic reference thermocouples that are of major importance in thermometry because of their excellent mechanical and chemical properties and high degree of homogeneity. These are, platinum vs. platinum + 13% rhodium (Type R), platinum vs. platinum + 10% rhodium (Type S), and platinum + 6% rhodium vs. platinum + 30% rhodium (Type B). It is the type S thermocouple that is used to define the International Practical Temperature Scale between 630.74 and 1064.43°C, although the type R thermocouple has slightly greater sensitivity. Both can be supplied either as high-quality Reference Grade or normal Standard Grade. They can be used in air or inert atmospheres for prolonged periods and in hydrogen with care, but they should not be used in a vacuum.

Finally, it should be emphasized that none of these thermocouples can be used below about 10–15 K, as the thermoelectric power of all of them is then extremely small. At those temperatures thermocouple thermometry is carried out by utilizing the "giant" thermopowers of noble metals containing transition metal ions, such as Au–Fe or Au–Co. A common thermocouple is Au+0.07 at.% Fe vs. chromel, which has a thermopower at 4.2 K of 13 μV deg^{-1}. Calibration tables for this and several similar thermocouples are readily available [61P1, 68S2, 68S3, 69S3, 72S1, 72S2].

6 | Dilute Magnetic Alloys

6.1 Introduction

We shall devote a separate chapter to dilute "magnetic" alloys because both experimental results and theoretical approaches set such alloys apart from those discussed in Chapter 2. The adjective "magnetic" has been set in quotation marks because we do not refer here to magnetic systems in the conventional sense. On the contrary, we specifically exclude from discussion alloys which exhibit a phase transition to a magnetically ordered state, ferromagnetic or antiferromagnetic. Our interest is in alloy systems of non-transition-metal solvents with transition-metal or rare-earth solutes; moreover, the solute concentration is restricted to small values, typically less than a few tenths of an atomic percent and always less than a few atomic percent. The adjective "magnetic" arises from two sources: (a) First and foremost, there is the observation that the magnetic susceptibility of many such systems displays properties consistent with the presence of localized and free (or nearly free) magnetic moments. (b) There is a growing consensus that spin fluctuations are important for transport properties in many of these systems despite the apparent lack of significant temperature dependence in their magnetic susceptibilities. The concepts that have proved useful in understanding the properties of these alloys are also applicable to some alloys which do not require free magnetic moments or spin fluctuations being associated with solute atoms. Thus we shall include all these systems in our discussion.

Historically, the unusual role of transition-metal impurities in influencing the transport properties of ordinary metals was first manifested in the

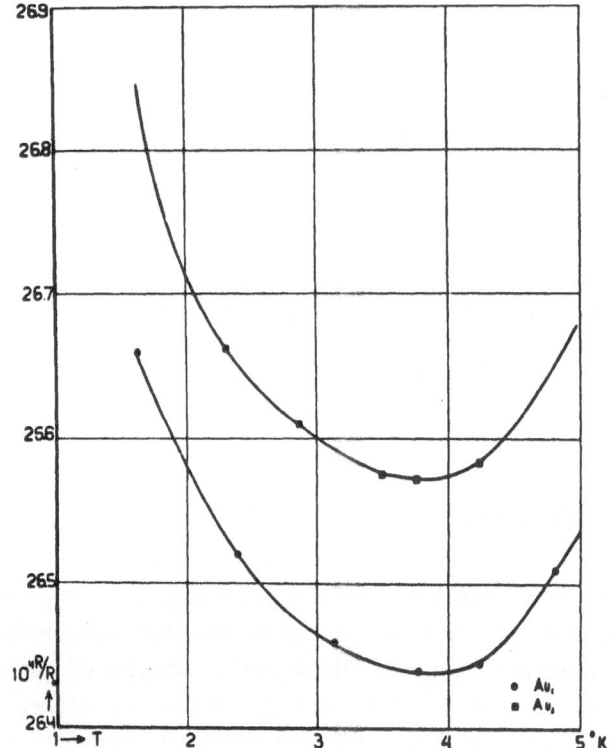

Fig. 6.1. Resistance vs. temperature for nominally pure gold. All resistance values are normalized by R_0, the value at 0°C (from [33D1]).

appearance of a low-temperature *resistivity minimum* in some nominally "pure" metals, notably copper and gold [33D1]. A typical curve of ρ vs. T for such a case is shown in Fig. 6.1. For many years following the initial observation of this phenomenon its cause remained a mystery. The low-temperature studies of the NRC group [53M1, 54M1] demonstrated that samples which display the resistance minimum also exhibit an anomalously large thermoelectric power at low temperatures, often referred to as a *giant thermopower*. Generally, as we have seen earlier, S near 4 K is between 0.01 and 0.1 μV/K; in these anomalous systems the thermopower attained values between 1 and 10 μV/K (see Fig. 6.2). Subsequent careful work revealed that the resistance as well as the thermopower anomalies could be traced to the presence of very small concentrations of transition-metal impurities, notably iron, in the nominally pure samples [60G1]. Moreover, magnetic susceptibility measurements on these alloy systems

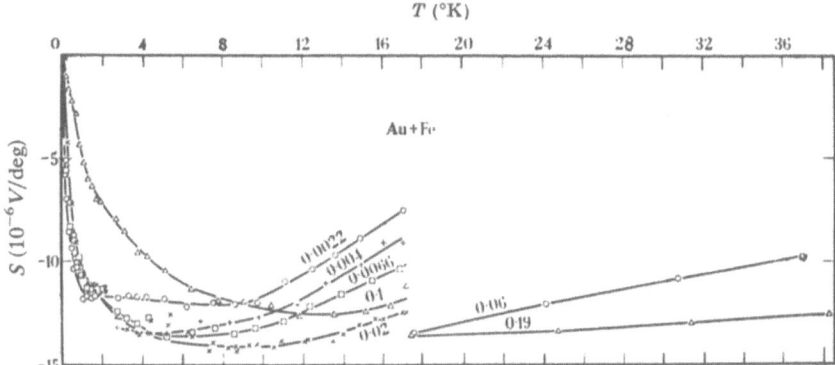

Fig. 6.2. Thermopower vs. temperature for AuFe. The number with each curve indicates the iron concentration in atomic percent (from [62M2]).

[74R1] also revealed a Curie–Weiss component at low temperatures, and the presence of this component led naturally to the concept of the impurity supporting a *local magnetic moment* (LMM). Further investigations have raised questions about the form of susceptibility behavior which uniquely documents the presence of LMM. However, a Curie–Weiss contribution in the susceptibility has proved to be an infallible indication that magnetic considerations are important, and it is generally accepted as an operational definition for the presence of LMM. In 1964, more than thirty years after the first observation of the resistance minimum, a theoretical explanation of the phenomenon based upon the magnetic properties of the impurity was published by Kondo [64K2]. The terms *Kondo systems* and *Kondo anomalies* have now become generally accepted. In a later paper, Kondo also addressed himself to the giant thermopower [65K1].

Although the effects of transition-metal impurities on normal metal transport properties are most dramatic when the impurities carry a local magnetic moment, even when such a moment appears to be absent the alloys of normal metals with transition-metal solutes manifest properties that set them apart from other alloy systems. For example, the residual resistivities are often quite large and generally do not conform to Linde's rule [68B1]. From the theoretical as well as the experimental viewpoint it is appropriate, therefore, to treat all transition-metal impurities in the same chapter, even though some do not display "magnetic" properties. In fact, there is an emerging consensus that the distinctions "magnetic" and "nonmagnetic" are not fundamental but are instead measures on an energy scale. Both types of alloys involve the same basic principles, but the location of the "non-magnetic" alloys on a far edge of the energy scale does not permit observation of magnetic effects at realizable temperatures. One unifying thread that makes the treatment of both magnetic and nonmagnetic transition-metal impurity systems meaningful is the concept of a *virtual bound state* (VBS) introduced by Friedel in 1956 [56F1].

6.2 The Virtual Bound State

Consider a metal, in the free-electron gas approximation, into which we introduce an impurity whose valence is greater than that of the host ions. The impurity *ion*, whose charge exceeds that of the host ions by ΔZ, presents an attractive potential to the free electrons; consequently, conduction electrons will tend to accumulate in its neighborhood so as to ensure electrical neutrality. If the impurity potential is very strong, a bound state may be formed well below the condution band edge, as shown in Fig. 6.3, which is reminiscent of the donor level in a semiconductor. In real space such a bound state localizes the electron about the impurity ion. For weaker potentials, i.e., smaller ΔZ, the energy of the bound state may lie within the conduction band, specifically, in the continuum. Such a state is not a truly bound, negative energy state but forms a VBS which is broadened by admixing with wave functions of neighboring energies. To quote Friedel:

It is useful to think of the bound state still existing, with a positive energy. But as it now has the same energy as an extended state, it will resonate with the lth spherical component, to build up two (extended) states of slightly different energies; these in turn will have the same energies as extended states with whom they will resonate, etc.

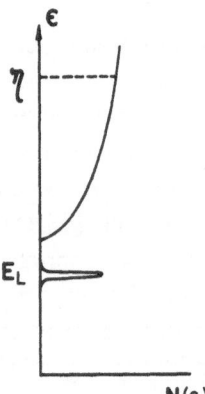

Fig. 6.3. Density of states vs. energy. The conduction band is represented by the usual parabolic function, and the bound state at the energy E_L is represented by a sharp peak.

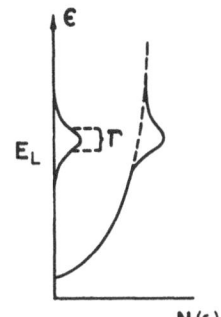

Fig. 6.4. Density of states vs. energy. The VBS, having an energy E_L and a half-width Γ, is depicted separately along the energy axis. Since this state is degenerate with the conduction band, the total density of states is the parabolic band plus the VBS, i.e., the outer solid line.

The VBS will manifest itself through a more or less narrow peak in the density of states at the energy of the VBS, as indicated schematically in Fig. 6.4. Since the electronic properties of a metal depend critically on the density of states at the Fermi energy, it is apparent that we can anticipate dramatic changes if the VBS energy coincides or is very close to the Fermi energy.

The picture is clarified and made more precise when regarded from the viewpoint of scattering theory using the phase-shift approach. The effect of the impurity potential is to produce, at large distance from the impurity, a phase shift δ_l in a spherical wave of angular momentum $\hbar\sqrt{l(l+1)}$. For a given potential the phase shifts are functions of the electron energy, as

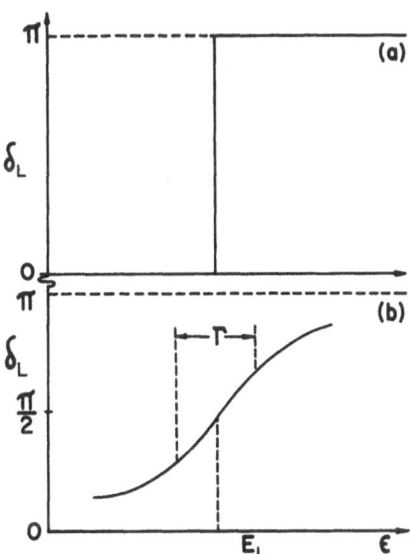

Fig. 6.5. Phase shift vs. energy. Part (a) depicts the abrupt change of δ_L by π, which occurs for a true bound state of energy E_L, and part (b) depicts the variation which occurs for a VBS.

shown in Fig. 6.5. For a true bound state with a well-defined energy E_L, the phase shift exhibits a discontinuity of π as ε goes from $\varepsilon < E_L$ to $\varepsilon > E_L$. For a VBS the change in δ_l with energy is nearly the same but occurs over a finite energy interval of order Γ, where Γ, the half-width of the VBS, is given by

$$\Gamma = 2\pi \overline{|\langle L|V|k\rangle|^2 N(\varepsilon)} \tag{6.1}$$

Here L denotes the atomic level of the impurity ion of angular momentum $\hbar\sqrt{L(L+1)}$, V is the potential coupling the impurity ion and the conduction electrons, k denotes the conduction electron state, and $N(\varepsilon)$ the conduction band density of states. Since Γ depends on the proportion of the lth component of the wave function for the conduction electrons, two qualitative conclusions hold: (a) For a given E_L, Γ decreases with increasing l; and (b) for a given l, Γ increases with increasing E_L. It turns out that for $l=0$ (s-states) Γ is so broad that the concept of a VBS loses its significance. Also for $l=1$ (p-states) Γ is still quite large generally, and the concept of a VBS proves most useful for d-states and f-states. Transition-metal impurities provide an obvious realization of the former states and rare-earth impurities a realization of both types.

In terms of scattering theory, a VBS results in resonant scattering at the energy $\varepsilon = E_L$, and hence a large scattering cross section. When E_L coincides with the Fermi enegy η, we can then expect an unusually large residual resistivity ρ_r. In addition, if Γ is small, δ_l and, therefore, also ρ_r will change rapidly with electron energy. Hence, we further anticipate that the diffusion thermopower S_d^r may show fairly large values. Moreover, since the sign of the derivative of ρ_r with energy depends on whether E_L is less than or greater than η (and is approximately zero when $E_L = \eta$), S_d^r may be of either sign, depending on the location of E_L.

6.2a   Electronic Properties for VBS Systems

Both ρ_r and S_d^r can be expressed in terms of the scattering phase shifts and their energy derivatives [65D3]. The phase shifts, in turn, must satisfy the Friedel sum

$$\Delta Z = \frac{2}{\pi} \sum_l (2l+1)\delta_l(\eta) \tag{6.2}$$

When E_L lies near the Fermi energy, the only phase shift of substantial magnitude is that for which $l = L$, and as a rough approximation one may then disregard all other δ_l in using the Friedel sum. In the following expression we shall assume that $L = 2$, that is, we are dealing with a transition-metal impurity. With the above approximation the phase shifts, the resistivity per atomic percent impurity, and the thermopower are given by

$$\delta_0 = 0 \qquad \delta_1 = 0 \qquad \delta_2(\eta) = \pi \Delta Z / 10 \tag{6.3}$$

$$\rho_r = \frac{\pi \hbar}{5 e^2 k_F} \sin^2 \delta_l(\eta) \tag{6.4}$$

$$S_d^r = \frac{\pi^2 k^2 T}{3 e \eta} \left[0.5 - 2 \left(\frac{d \delta_2}{d \ln \varepsilon} \right)_\eta \cot \delta_2(\eta) \right] \tag{6.5}$$

Since the phase shift $\delta_2(\varepsilon)$ is related to the energies ε and E_L and also to the width of the VBS by

$$\tan \delta_2(\varepsilon) = \frac{\Gamma}{2(E_L - \varepsilon)} \tag{6.6}$$

the expression for S_d^r can be rewritten in terms of Γ, i.e.,

$$S_d^r = \frac{\pi^2 k^2 T}{3 e \eta} \left\{ 0.5 - \frac{2 \eta}{\Gamma} \sin \left[2 \delta_2(\eta) \right] \right\} \tag{6.7}$$

Equation (6.7) is essentially the expression given by Boato and Vig [67B5] except for the term 0.5, which they omitted and which is usually quite small in comparison with the second term in the braces.

The appearance of a VBS near the Fermi energy exerts an influence on the equilibrium electronic properties as well as on transport. For example, since the electronic specific heat of a metal is proportional to $N(\eta)$ [65D3], we expect an additional contribution to C_e due to the presence of the VBS given by

$$\Delta C_e = \frac{\pi k^2 T}{30} \left(\frac{d \delta_2(\varepsilon)}{d \varepsilon} \right)_\eta \tag{6.8}$$

where ΔC_e is the contribution per atomic percent impurity. Assuming a Lorentzian shape for the VBS and using Eq. (6.6), we obtain

$$\Delta C_e = \frac{\pi k^2 T}{15} \frac{\Gamma}{(E_2 - \eta)^2 + \Gamma^2} \tag{6.9}$$

We shall note here the following general features of the above results. According to Eq. (6.3), $\delta_2(\eta)$ assumes the value $\pi/2$ when $\Delta Z = 5$, and Eq.

(6.6) indicates that the energy of the VBS equals the Fermi energy irrespective of the value for Γ. When this occurs, $\sin \delta_2 = 1$ and ρ_r attains its maximum value and, according to Eq. (6.9), ΔC_e should then show a maximum if Γ were independent of E_L. From Eq. (6.7) the impurity thermopower should then be relatively small ($\sin 2\delta_2 = 0$), and S'_d should be positive for $\Delta Z < 5$, and negative for $\Delta Z > 5$.

It might seem at first that the extension of these ideas to rare-earth impurities requires no more than setting $L = 3$ and retracing the above logic. However, this initial impression is incorrect as one may see by considering a rare-earth impurity in a noble-metal host. The rare-earth impurity has outer electrons in the configuration $4f^n 6s^2 5d^1$; therefore, ΔZ includes f-state *plus* s- and d-state electrons. In the preceding paragraph d-state electrons alone were the source of VBS, and an understanding of the role of these electrons for rare-earth impurities is essential. Anticipating experimental results somewhat and assuming that the f-state electrons and the d-state electrons have independent roles, one may imagine that the f-states are very narrow with energies well removed from the Fermi level and that the d-state electrons behave as previously depicted, thus dominating the observed transport properties. This model will prove quite satisfactory with the exception of instances when the f subshell is either almost empty or almost filled for one or both spin orientations. In these instances the behavior is more complex.

6.2b   *Survey of Experimental Results for VBS Systems*

Let us now compare some of the predictions of the VBS model with observations. Most of the measurements of interest have been made on aluminum alloys. Boato *et al.* [66B2] also studied some Zn, In, and Sn alloys, but data on these systems are largely limited to residual resistance measurements. Many alloys of the noble metals with transition elements exhibit resistance minima indicative of an exchange split VBS and, thus, must be analyzed in terms of Kondo's theory. CuNi and a few silver and gold alloys do not exhibit local magnetic moments, and we shall tabulate the relevant results.

The residual resistivities of the aluminum alloy system are shown in Fig. 6.6. Evidently, as the number of d-electrons increases, the residual resistivity passes through a maximum near AlCr. The systematic trend and the location of the ρ_r maximum near AlCr, with atomic Cr having $5d$-electrons, are consistent with Eqs. (6.3) and (6.4) and appear to indicate that the

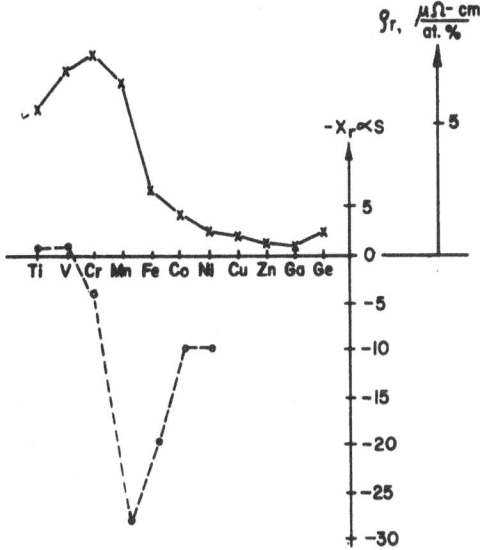

Fig. 6.6. Residual resistivity and impurity thermopower for aluminum alloys. The solid line connects the residual resistivity values (from [65D3]), and the dashed line connects the experimental x_r values [see Eq. (2.42)].

condition $E_L \simeq \eta$ is met near Cr. Two factors prevent this interpretation from being wholly convincing. First, Al is a valence three metal. The phase shifts given in Eq. (6.3) have ΔZ equaling the number of d-electrons for a monovalent host but require nonzero δ_0 and δ_1 and/or some difference between ΔZ and the number of d-electrons for a trivalent host. Second, the magnitude of ρ_r is nearly twice as large as predicted by Eq. (6.4). Such a discrepancy is perhaps not too distressing in view of the use of the free-electron model, a model known to give similar factors for ρ_r in alloys of Al with non-transition-metal impurities. It would, of course, be desirable to perform similar measurements for alloys based upon the alkali metals; for these the free-electron model should be an excellent approximation. Unfortunately, nature is most uncooperative in that transition metals are essentially insoluble in the alkalis. Boato *et al.* point out that the ratio of $\rho_r^{exp}/\rho_r^{theo}$ is almost constant for the entire sequence of aluminum alloys and suggest that the VBS arguments with ΔZ approximately equal to the number of d-electrons are fundamentally correct.

Measurements of S_d' for these alloys lend further support to the VBS model. As we pointed out above, as ΔZ is varied, S_d' should change sign near the value of ΔZ for which ρ_r is a maximum. Indeed, S_d'/T is small but positive for AlTi and AlV, is negative for AlCr, and attains a negative extremum for AlMn or AlFe, as shown in Fig. 6.6. According to Eq. (6.7), S_d' depends on both $\delta_2(\eta)$ and Γ, but if Γ were constant for the entire alloy system (a rather poor assumption), the negative extremum should come when $\delta_2 = 3\pi/4$, i.e., when $\Delta Z = 7.5$. Associating ΔZ with the number of d-electrons, $\Delta Z = 7$ corresponds to Fe, and the experimental results again appear to provide a qualitative confirmation of the VBS model.

It is obvious from Eqs. (6.4), (6.6), and (6.9) that measurement of ρ_r and ΔC_e should permit evaluation of the two parameters of the model, E_2 and Γ. Similarly, E_2 and Γ could also be determined from measurements of ρ_r and S_d'. A comparison of these parameters deduced from two such sets of data is given in Table 6.1, and it is apparent that there are substantial disagreements. Again, such quantitative discrepancies are not surprising in view of

Table 6.1. VBS Parameters for Dilute Aluminum Alloys[a]

Impurity	Γ (eV)	$E_2 - \eta$ (eV)	Γ (eV)	$E_2 - \eta$ (eV)	Ref.
Ti	23.4	10.0	–	–	–
V	20.9	6.3	0.62	0.1	[69A1]
			0.62	0.1	[74G8]
Cr	(b)	~0	0.46	~0	[69A1]
			0.46	~0	[74G8]
Mn	0.36	−0.04	0.34	~0	[69A1]
	0.54[c]	–	1.94	~0	[74G8]
			2 to 3[d]		[67R2]
Fe	1.17	−0.43	–	–	–
	1.0[c]				
Co	2.57	−1.16	–	–	–
Ni	2.0	−2.11	–	–	–

[a] Entries in columns 2 and 3 are based upon resistivity and thermopower data and were calculated using Eq. (6.4), Eq. (6.7), and experimental data. The data were taken from [74R1], and the right-hand side of Eq. (6.4) was multiplied by the empirical factor of 1.55 given in this reference. The very large Γ values for AlTi and AlV are surely an artifact of using this procedure with small thermopower values. Entries in columns 4 and 5 are based upon resistivity and specific heat data and were taken from the references in column 6.
[b] No estimate possible since $\delta_2 = 90°$. If the empirical scaling factor for resistivity were changed, an estimate becomes possible.
[c] Values obtained in [67B5] using a slightly different treatment of resistivity.
[d] This value actually comes from an analysis including changes in superconducting properties. It is included here because it supports the AlMn value from [74G8] and it provides some indication of the inconsistency in values which occurs when numerous properties are considered. Table III in [69A1] contains a related comparison in terms of the number of d-electrons deduced from various properties.

Table 6.2. VBS Parameters for $\underline{Cu}Ni^{a}$

Method	$\Gamma/2$ (eV)	$E_2 - \eta$ (eV)	Concentration of Ni (at.%)
Specific heat	0.3	-0.93	10
Thermopower	0.25 ± 0.10	-0.70 ± 0.05	<1
UPS	0.55	-0.95	13
Optical	0.27 ± 0.02	-0.75 ± 0.02	1
XPS	0.35 ± 0.10	-0.8 ± 0.1	0
CPA	0.23	-0.8	13
Theory	0.3	–	–

[a] From [73H2]. UPS and XPS denote ultraviolet and X-ray photoelectron spectroscopy, and CPA denotes coherent potential approximation. References for the individual studies are available in the original source.

Table 6.3. Impurity Thermopower for VBS Alloys[a]

Alloy	Temperature range	$-x_r$	Ref.
$\underline{Al}Ti$	6–10	0.47	[67B5]
$\underline{Al}V$	0.6–300	0.47	[67B5], [74B1]
$\underline{Al}Cr$	0.6–300	-3.8	[67B5], [74B1]
$\underline{Al}Mn$	0.6–300	-28.0	[67B5], [74B1]
$\underline{Al}Fe$	0.6–300	-19.4	[67B5], [74B1]
$\underline{Al}Co$	1.5–300	-9.5 ± 1.0	[74C3]
$\underline{Al}Ni$	1.5–300	-9.5 ± 1.5	[74C3]
$\underline{Cu}Ti$	77–300	~6.5	[69B2]
$\underline{Cu}V$	77–300	~5.9	[69B2]
$\underline{Cu}Ni$	5–25	-17.6	[68F4]
$\underline{Cu}Pd$	77–400	-2.8	[56O1]
$\underline{Cu}Pt$	77–400	-3.3	[56O1]
$\underline{Au}V$	2–20	18.0	[67D4]
$\underline{Ag}Pd$	77–300	-3.2 to -4.8	[65S1]
$\underline{Ag}Pd$	77–400	-4.2	[56O1]
$\underline{Ag}Pt$	77–400	-3.6	[56O1]
$\underline{Ag}Sc$	1.6–6	4.5	[72B6]
$\underline{Ag}Y$	1.6–6	1.6	[72B6]
$\underline{Ag}Nd$	1.6–6	0.45	[72B6]
$\underline{Ag}Sm$	1.6–6	1.1	[72B6]
$\underline{Ag}Gd$	77–300	3.4	[71F3]
$\underline{Ag}Ho$	77–300	2.7	[71F3]
$\underline{Ag}Tm$	77–300	3.1	[71F3]
$\underline{Ag}Lu$	1.6–6	2.4	[72B6]

[a] The typical x_r values for non-transition-metal impurities range from 0.14 to 1.73 (see Chapter 2).

the numerous simplifying approximations, such as neglect of $l = 0$ and $l = 1$ phase shifts and the arbitrary assumption of a symmetrically broadened, Lorentzian VBS.

Rather better agreement for $E_2 - \eta$ and Γ from diverse measurements and theoretical calculations is obtained for CuNi, as indicated in Table 6.2. It is unfortunate that exchange splitting of the VBS in CuMn and CuFe precludes similar comparison for the entire system of copper–transition metal alloys.

We conclude this section with Table 6.3, which gives values of x_r (i.e., $\cdots - (d \ln \rho_r / d \ln \varepsilon)_\eta$, see Section 2.7a) for a number of VBS alloys. The important point here is that in almost every instance x_r is substantially greater (in magnitude) than the typical values that obtain for non-transition-metal solutes given in Table 2.2. The data for the rare-earth impurities in Ag as well as the data for AgSc and AgY are collected at the end of Table 6.3. Sc and Y have outer electron configurations of $3d^1 4s^2$ and $4d^1 5s^2$, respectively, and thus they resemble the outer electron configuration $5d^1 6s^2$ which typifies the rare-earth series. The residual resistivities for all these systems do not differ by more than a factor of two and, with the exception of AgNd and possibly AgSm, the thermoelectric parameters listed in Table 6.3 are comparable. It should be noted that all the values of x_r are negative and there is no significant correlation between x_r and the number of f-state electrons. This behavior appears to differ significantly from that for transition-metal impurities in Al. The assumption that the f-states are well below the Fermi level while the d-states form VBS which dominate transport properties predicts that all the rare-earth impurities should produce a similar VBS structure which is independent of the number of $4f$-electrons, and which is typical of the initial elements in a transition-metal sequence. Such an explanation appears consistent with all data in Table 6.3. Experimental indications for deviation from the preceding pattern are too limited to permit quantitative considerations, but they always occur when the impurity undergoes a change in its valence [67G3, 70A2].

6.3 Kondo Alloys

We now turn to those alloy systems that display obvious magnetic properties and striking anomalies in their transport properties. In these systems, such as AuFe, CuFe, ZnMn, ZnCr, and many others, there is ample evidence that the VBS for the two possible spin directions appears at

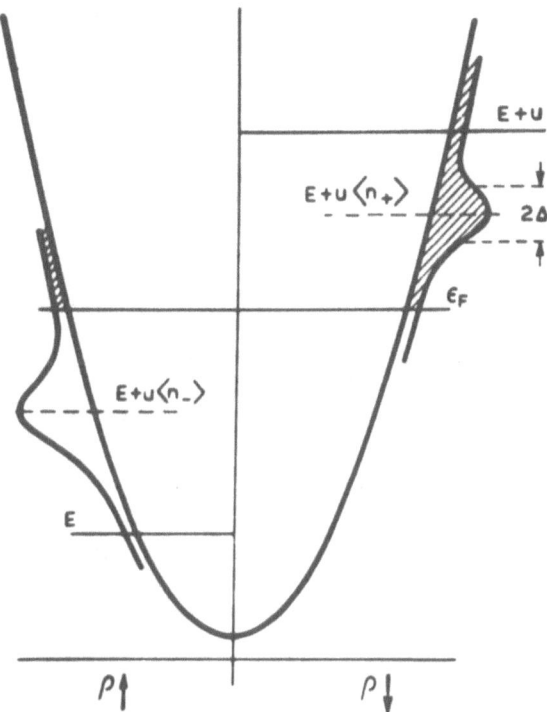

Fig. 6.7. Density of states for an exchange-split VBS. Virtual d levels of widths 2Δ appear at $E + u\langle n_-\rangle$ and $E + u\langle n_+\rangle$, where E is the energy of the atomic d state. The number of electrons in these levels is computed from the area of the *unshaded* portion, below the Fermi energy.

different energies, as shown schematically in Fig. 6.7. As a result of exchange interaction, a lowering of energy occurs when the d-electrons have their spins aligned (as far as is consistent with the exclusion principle), resulting in a net spin and magnetic moment at the impurity site. The conditions for formation of a local moment have been investigated in some detail, and as a rough guide local moments will appear when $E_L \sim \eta$ and the exchange energy is greater than Γ.

One consequence of a spin-split VBS is that as ΔZ increases, two density-of-states extrema will traverse the Fermi energy. As a result, one may expect that a plot of ρ_r vs. atomic number will exhibit two extrema, as shown in Fig. 6.8, in contrast to the single maximum of Fig. 6.6. Another

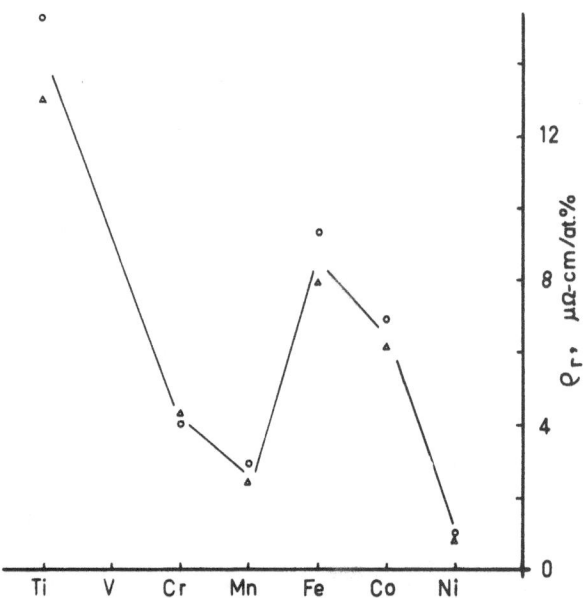

Fig. 6.8. Residual resistivity of dilute copper-based (O) and gold-based (△) alloys. The straight lines indicate the general trend, which contrasts sharply with that shown in Figure 6.6.

consequence of spin splitting of the VBS is that with regard to the thermopower the exchange scattering of conduction electrons assumes a dominant role and tends to obscure effects associated with resonant potential scattering, which was discussed in the preceding sections.

The most direct evidence for the presence of local magnetic moments comes from measurements of magnetic susceptibility. For a degenerate Fermi gas, χ assumes a constant (temperature-independent) value, the sum of the Pauli spin paramagnetic and Landau diamagnetic contributions [53W1]. Local magnetic moments add a Curie–Weiss term of the form

$$\chi = \frac{\mu_e^2}{3k(T+T_c)} \tag{6.10}$$

where μ_e is the effective moment, and T_c is a Curie–Weiss temperature. Since the emphasis of this book is on transport, we shall not dwell further on the magnetic properties except to reiterate that they give some confirmation to the concepts discussed below. For an up-to-date review of the subject, the reader is referred to *Magnetism*, Vol. V [73S1].

The phenomena of interest to us and which theory should explain are the resistivity minimum and the associated giant thermopower. Unfortunately, the theory is extremely difficult and, moreover, several different though somewhat complementary avenues have been pursued. As Suhl has stated, "—successive waves of theorists have attacked this problem but, although substantial inroads have been made, it still represents a very considerable challenge" [71S2]. In view of the many still unresolved problems of a fundamental nature, any detailed discussion of predictions for thermoelectric behavior is unwarranted, and we shall give only a broad, cursory overview. In so doing, we cannot nor do we attempt to do justice to the many important contributions to this field. Again we must refer the reader to the articles in *Magnetism*, Vol. V, for a good exposition of the problems and their partial solutions.

6.3a *Theory of the Kondo Effect*

The earliest approach, used by Kondo [64K2, 65K1], postulated an exchange interaction of the form $\mathcal{H} = -2J\mathbf{s}_i \cdot \mathbf{S}_j$, where $\mathbf{s}_i$ refers to the spin of the conduction electron, and $\mathbf{S}_j$ to the spin of the d-electrons at the impurity site. The effect of this s–d interaction on the electron relaxation time was then calculated using perturbation theory. Kondo found that the resistivity (excluding the contribution due to electron-phonon scattering) diverges logarithmically as $T \to o$. Specifically, he found that

$$\rho_r = \rho_r^0 \left[1 + 2JN(\eta) \ln \frac{T}{T_K} \right] \tag{6.11}$$

where, for our purposes, T_K (the Kondo temperature) is simply a parameter in the theory. If J is negative, corresponding to "antiferromagnetic" exchange coupling, ρ_r will increase with decreasing temperature. Since the ideal resistivity, due to electron–phonon scattering, has the opposite trend, the total measured resistivity must display a minimum at some intermediate temperature.

Although Kondo's result provided an explanation for the resistivity minimum, it also demonstrated the questionable validity of perturbation theory when applied to this problem and thus raised far more questions than it answered. Following Kondo's work, others have attacked the problem using various sophisticated techniques, such as dispersion theory [67S5] and Green's function methods [65N1, 67N1]. One of the conclusions of these theories is that as the temperature is reduced toward $T = 0$ K, the localized

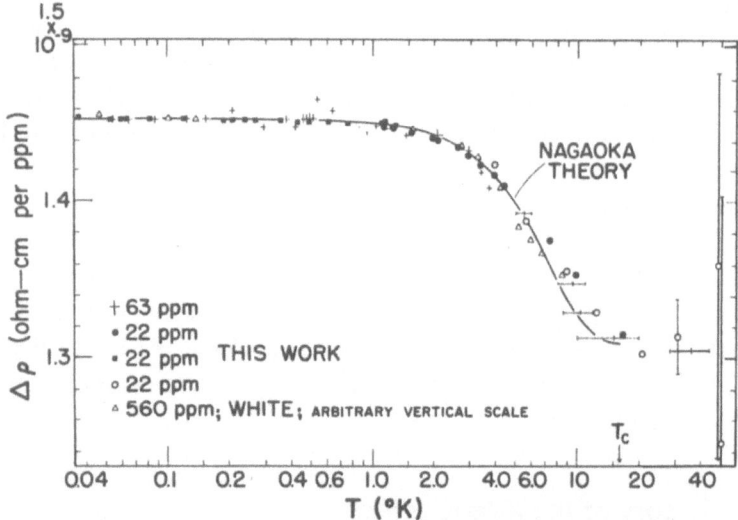

Fig. 6.9. Low-temperature resistivity of several dilute C̲u̲Fe alloys are shown to lie on a single curve which saturates at low temperature. The large error bars at higher temperature are associated with phonon contributions to resistivity and the difficulties in separating the impurity contributions from the total resistivity (from [67D3]).

magnetic moment will be replaced by a spin-compensated ground state of the system with vanishing magnetic moment. The existence of such a ground state resolves the difficulty posed by the divergence of Kondo's result. As shown in Fig. 6.9, the resistivity, of course, does not increase indefinitely with diminishing temperature but approaches, instead, a constant value. Within the single phase shift context of Eq. (6.3), the residual resistivity has a maximum value which is uniquely related to ΔZ. This maximum value is called the unitarity limit, and it is generally believed that this limit is approached in all Kondo systems at temperatures well below T_K.

In view of the diverse theoretical calculations for resistivity, for magnetic susceptibility, and for thermopower as well as superconducting properties, it is difficult to provide an unequivocal definition of the Kondo temperature. From the vantage point of Suhl's approach, the system (conduction electron gas plus transition-metal impurity) always condenses into a spin-compensated ground state (which has sometimes been viewed as a quasi-singlet state) at sufficiently low temperatures. The Kondo–Suhl temperature is, therefore, that temperature at which thermal fluctuations break up the ground state and excite the system into a magnetic excited state. In

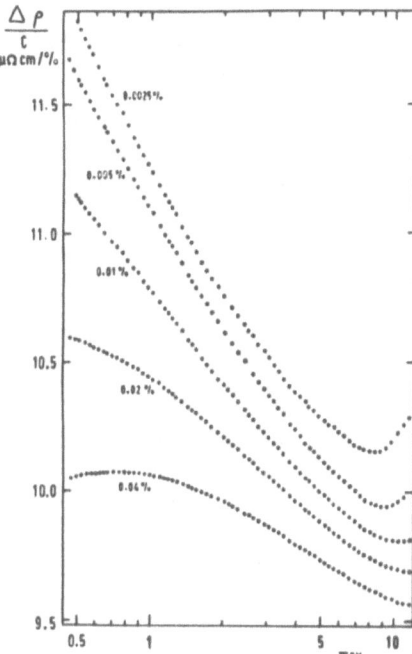

Fig. 6.10. The resistivity of A̲uFe alloys at low temperature (from [70L2]). In contrast to the C̲uFe system, shown in Fig. 6.9, no single universal curve is obtained. The different behavior is presumably due to impurity–impurity interactions.

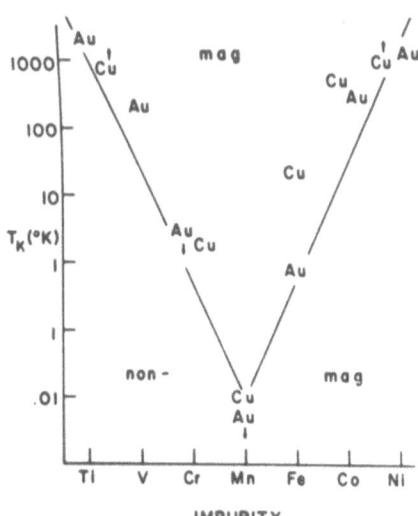

Fig. 6.11. Kondo temperatures for copper and gold alloys (from Daybell [73S1]).

this context all alloys of transition-metal impurities in non-transition-metal hosts are potential Kondo systems, and failure to observe a Kondo anomaly in alloys such as A̲lFe or C̲uNi may be ascribed to extremely high values of T_K. Taking, perhaps, a more pragmatic approach, we can deduce a Kondo temperature whenever the "residual" resistivity, that is, $\Delta\rho(T) \equiv \rho - \rho_i$, follows a logarithmic temperature dependence over at least some restricted temperature range, as in the case of some gold alloys shown in Fig. 6.10. T_K is estimated as the temperature at which $\Delta\rho = \frac{1}{2}[\Delta\rho(0) - \Delta\rho(\infty)]$. In making this estimate, it is essential to exercise care in separating the phonon-scattering contribution and to study enough concentrations to ensure the logarithmic slope has become independent of impurity concentration.

Since T_K depends exponentially on the exchange coupling constant, small changes in J have a profound influence on T_K. Figure 6.11 shows T_K for gold and copper alloy systems, and Kondo temperatures for other alloys are listed in Table 6.4. It is apparent from Fig. 6.11 that the logarithm of T_K

Table 6.4. Thermopower Summary for Kondo Systems[a]

Host	Impurity	Peak magnitude (μV/K)	Peak sign	T_k (K)
Mg	Mn	3.6	−	<30
Al	Mn	<1	−	>80
Cu	Ti	−	+?	−
	V	−	+?	−
	Cr	0.2?	−?	5?
	Fe	16	−	22
	Co	~15	−	240
	Ni	>15	−	>1000
Zn	Mn	5	−	<20
Ag	Mn	−	−?	<4
	Fe	7	−	<10
Au	V	5	−	240
	Cr	>2	+	~4
	Mn	−	+?	<9
	Fe	14	−	<4
	Co	40	−	270
	Ni	16	−	550

[a] From Daybell in [73S1]. These systems all exhibit some extremal values in thermopower which are clearly not associated with phonon drag. The values in this table are inferred from thermopower data and are semiquantitative at best. Several systems listed here, notably A̲lMn, C̲uTi, C̲uV, and C̲uNi and probably C̲uCo, AuCo, and A̲uNi, are often classified as VBS systems. The extrema in these systems occur at high temperatures and may indicate that T_K is being approached or may be the natural result of competition between a positive host and a negative impurity contribution.

varies roughly inversely with the number of unpaired $3d$-electrons on the impurity.

6.3b *Thermoelectric Power of Kondo Alloys*

Turning now to the topic of primary interest to us, Fig. 6.2 shows the thermoelectric power of one Kondo system, and peak values of the thermopower for others are given in Table 6.4. The remarkable fact about these results is, as we mentioned earlier, the very large value of S at quite low temperatures. Generally, S for dilute Kondo alloys exhibits an extremum near T_K, and thus Kondo temperatures can be inferred from thermopower data.

Theoretical studies of the thermopower of Kondo systems were undertaken by Kondo [65K1], Suhl and Wong [67S4], Fischer [67F3, 69F2, 71F2], Maki [69M2], Chow and Everts [69C2], and Weiner and Beal-Monod [70W1]. Kondo's treatment, based on a perturbation approach, while providing some understanding of the origin of giant thermopowers, was inadequate because terms were omitted which proved to be of some consequence. These terms were subsequently included in the work of Fischer [71F2]. Suhl and Wong [67S4] presented their results, obtained by extensive numerical calculation, in graphic form; their paper contains several sets of curves showing the thermopower as a function of temperature for a large selection of values of the relevant parameters, such as strength of potential scattering, exchange coupling, and a parameter characterizing the range of interaction. The curves replicate many of the features of the experimental results. Fischer, who employed a Green's function technique, was able to arrive at an analytic expression for S'_d. Similar but not identical results were deduced by Maki and by Chow and Everts. Maki's result is quoted as a convenient illustration of the complexity and the parameters involved in analytical expressions:

$$S'_d = \frac{\pi k}{2e} \frac{\sin 2\delta_v}{1 - \cos 2\delta_v (\ln (T/T_K)/\{[\ln (T/T_K)]^2 + \pi^2 S'(S'+1)\}^{1/2})}$$
$$\times \frac{\pi^2 S'(S'+1)}{\{[\ln (T/T_K)]^2 + \pi^2 S'(S'+1)\}^{3/2}}$$

(6.12)

Here S' is the impurity spin and δ_v is a single-phase shift used to characterize potential scattering.

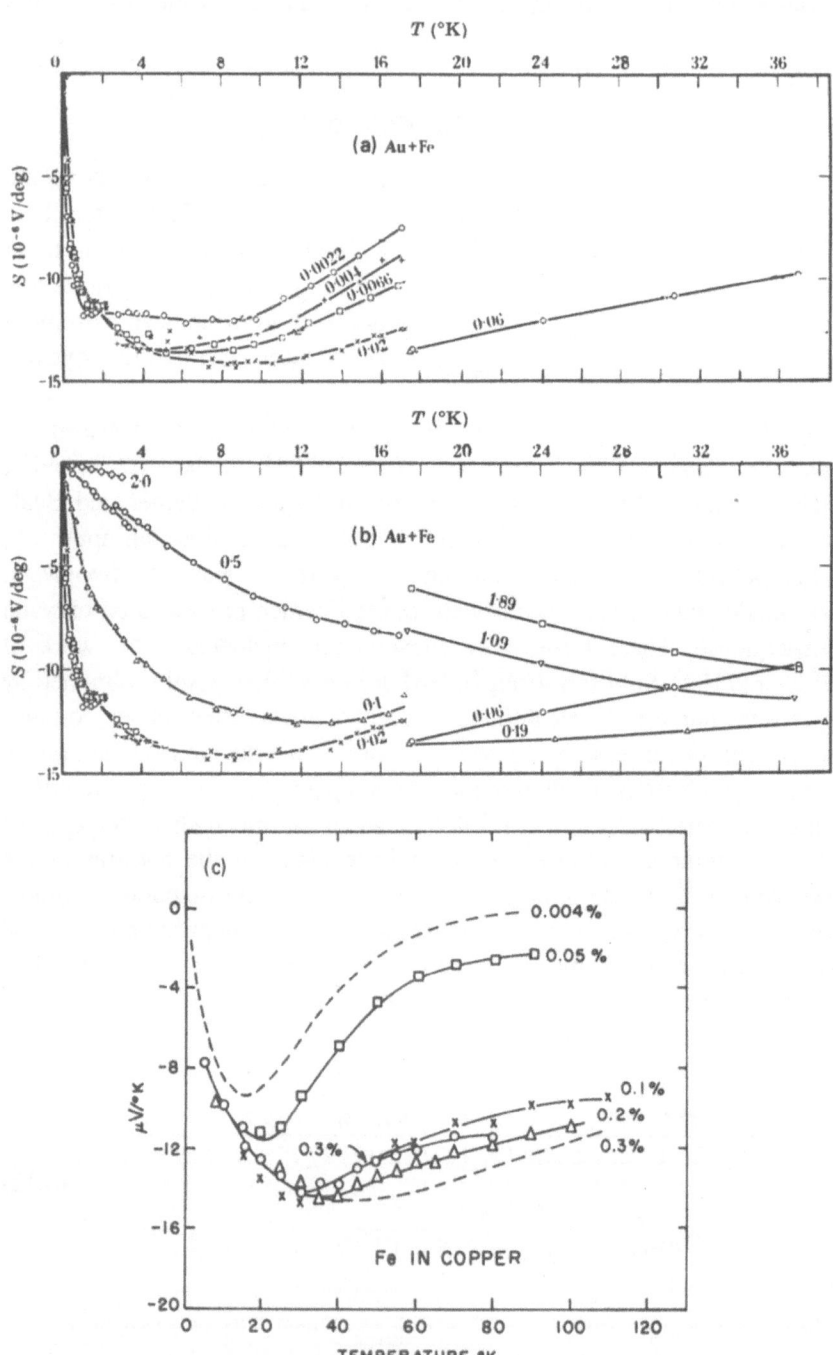

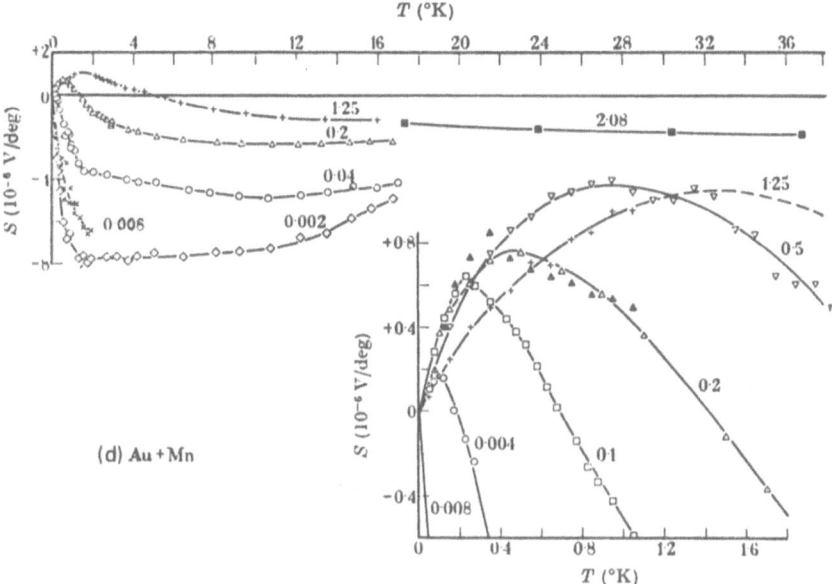

Fig. 6.12. Thermopower in some classic Kondo systems. (a, b) A̱u̱Fe (from [62M2]); (c) C̱u̱Fe (from [63C1]); (d) A̱u̱Mn (from [62M2]). The numbers along the curves give the impurity concentration in atomic percent.

Despite their differences all of the results do have certain common features. First, potential as well as exchange scattering is required to produce a giant thermopower; this is consistent with conclusions reached initially by de Vroomen and Potters [61D2] using classical arguments (see also Section 5.2c). Second, the thermopower S'_d is independent of the concentration of the magnetic impurities. The observed dependence on concentration for dilute alloys is, in all the theories, simply the result of the resistivity weighting factors [see Eq. (2.50)]. Third, the sign of the thermopower is determined by a number of factors, not simply by the sign of the charge carriers; this is, of course, not peculiar to the magnetic impurity system as we saw in Chapter 2. Although all of the results reflect the dominant experimental features, an extremum of the order of $0.3k/e$ at or near T_K, there are also significant differences which underscore the fact that the current state of the theory is not yet satisfactory.

As we turn to a brief survey of the experimental results, we show in Fig. 6.12 some data for a variety of "classic" Kondo systems. The curves in Figs. 6.12(a) and 6.12(c) follow the expected pattern: At low temperatures the thermopower is independent of concentration; an extremum is reached at a temperature which increases slightly as the concentration increases as a result of the resistivity weighting factor. The results in Figs. 6.12(b) and 6.12(d), however, show important quantitative and qualitative departures from the "standard" pattern; in Fig. 6.12(b) the thermopower extremum diminishes with increasing iron concentration; Fig. 6.12(d) the negative extremum is preceded by a low-temperature positive maximum whose value and temperature are sensitive functions of manganese concentration.

These and corresponding departures from the normal Kondo effect pattern in other properties of various alloy systems are generally attributed to impurity–impurity interactions which have, heretofore, been neglected. From perturbation theory one finds that the s–d exchange interaction leads to a conduction electron spin density which has the spatial dependence $\cos(2k_f r)/r^3$. Hence, one should include in the Hamiltonian a term of the Ruderman–Kittel–Kasuya–Yoshida form

$$\mathcal{H}_{RKKY} \propto \frac{J^2 S_1 \cdot S_2 \cos(2k_f r)}{r^3}$$

which can be expected to modify the behavior of the various properties associated with the Kondo effect. In this expression S_1 and S_2 are two impurity spins separated by a distance r. It has been suggested on empirical as well as theoretical grounds [71S3, 73R3] that significant interaction effects are to be expected when the impurity concentration exceeds approximately 50 ppm times T_K. It is important to emphasize here that this is a "rule of thumb," a very rough guide at best, and interaction effects judged from the basis of a single, specific property may appear at significantly different concentrations.

In regard to the thermopower, an indication of interaction effects can be deduced from curves such as those in Figs. 6.12(b) and 6.12(d). The existence of low-temperature peaks and sign reversals as a function of temperature indicates interactions are present, and Matho and Beal-Monod [74M1] use a pair description for the impurity–impurity influence upon the thermopower. This approach does not predict a specific concentration at which impurity–impurity interactions become significant. It predicts, instead, a thermoelectric behavior which scales with $\bar{n}$, where $\bar{n}$ is a parameter characterizing the relative number of impurity sites experiencing a magnetic environment. $\bar{n}$ depends upon temperature, the strength of

RKKY coupling, and impurity concentration; $\bar{n} \ll 1$ corresponds to the isolated impurity situation. This model reproduces the peaks and the sign reversals depicted in Fig. 6.12(d).

Indications of interaction effects can also be deduced from the behavior of the thermopower in a magnetic field *H*. Kondo [65K1] considered the effect of *H* on S_d' and concluded that the field, by impeding the internal degree of freedom for the impurity, should reduce the absolute value of the thermopower. Weiner and Beal-Monod [70W1], who included fourth-order terms in their perturbation calculation, although arriving at results at variance with Kondo's less elaborate calculation, came to the same general conclusion. The experimental results of Huntley and Walker [69H1] and of Berman and Kopp [71B2] on dilute AuFe alloys showed that S_d' at first increased with *H* and showed a diminution only at higher fields. The behavior was interpreted as indicative of interaction effects. Based on their data, Huntley and Walker suggested that interaction effects in this system may appear at Fe concentrations as low as about 1 ppm. Use of the earlier "rule of thumb" suggests an onset of significant interaction effects at impurity levels near 150 ppm, approximately two orders of magnitude greater than the value suggested by Huntley and Walker. This discrepancy plus the observation that no single concentration emerges from the treatment of Matho and Beal-Monod provide clear illustrations of the warning concerning the precision of this "rule of thumb."

Since T_K for a given transition-metal impurity depends critically on the host metal, several studies have been devoted to investigations of Kondo anomalies in ternary alloys, such as $Cu_{1-x}Al_x$Fe [72S3], $Cu_{1-x}Zn_x$Fe [68C1], $Cu_{1-x}Pd_x$Fe [69V2], and $Au_{1-x}Cu_x$V [72S4]. For example, the fact that CuFe shows a Kondo effect whereas AlFe does not suggests that T_K in AlFe is so high that the spin-compensated ground state persists at all normal temperatures in the latter system. Thus, for the alloy $Cu_{1-x}Al_x$Fe, one may expect that T_K would increase with increasing *x*. The results on this system [72S3] and on $Au_{1-x}Cu_x$V [72S4] seem to confirm this expectation. Unfortunately, there are various problems that plague such studies of ternary systems. For instance, Mössbauer measurements by Window [71W1] indicate that in CuAlFe alloys iron locates preferentially near an aluminum ion. In the CuPdFe system the thermopower changes sign at a Pd concentration between 20 and 30%, which was interpreted as indicative of a sign change in the exchange coupling [69V2]; however, in this same concentration range CuPd exhibits an order–disorder transition which also influences the transport properties.

6.4   Spin-Fluctuation Models

The use of spin-fluctuation models to calculate thermoelectric behavior is in an initial stage and, therefore, an accurate evaluation of this approach is not possible. In fact, most calculations of the effects of spin fluctuations upon transport properties have been concerned with resistivity and have ignored thermopower despite some rather striking effects in the latter property (see Section 5.2c). Fortunately, two recent studies [74F2, 74Z1] have considered the thermopower in some detail and provide an indication of the potentialities and scope of spin-fluctuation models.

Zlatić and Rivier [74Z1] have considered transition metals in aluminum. They begin by noting that the systematic patterns in ρ_r and S'_d predicted by a VBS model receive only very qualitative confirmation from experimental data; the strong asymmetry in x_r (see Figure 6.6 and Table 6.3) is deemed particularly significant. They introduce values of δ_0 and δ_1 to provide a better fit to ρ_r data (see the discussion in Section 6.2b) and then consider the effect of local spin fluctuations (LSF) upon thermopower. LSFs prove to be important in determining the expression for the resonant phase shift δ_2, and a new parameter M, the inverse of the LSF lifetime, enters the problem. For AlMn and AlCr the VBS still occurs at the Fermi energy (see Table 6.1), which necessitates $\delta_2 = \pi/2$ and thus ρ_r is unaffected. However, the energy derivative of δ_2 for these two alloys is enhanced to $1/M$, and this enhancement causes large increases in the predicted values for x_r. The prediction for AlMn is in good agreement with experiment, but for AlCr the prediction is too large by more than a factor of 2. No prediction is given for AlFe.

Fischer [74F2] uses an entirely different procedure. He calculates the dynamic susceptibility due to LSF, using the random phase approximation and a generalized model which permits modifying hopping of an electron from or into an impurity site as compared to hopping between lattice sites of the host. When a strong reduction of the hopping from or into an impurity site occurs, a sharp peak in the local density of state appears. This peak indicates well-localized electrons at the impurity site and is very similar to the VBS result. Fischer then restricts the problem to hosts having very weak exchange enhancement and calculates the transport properties.

The final results for the thermopower cannot be expressed in analytical form and they depend upon three parameters: (a) $|V|$, which is a measure of self-energy; (b) Δ, which is a function of the host bandwidth and the reduction in hopping terms; and (c) T_S, which is the LSF temperature. It

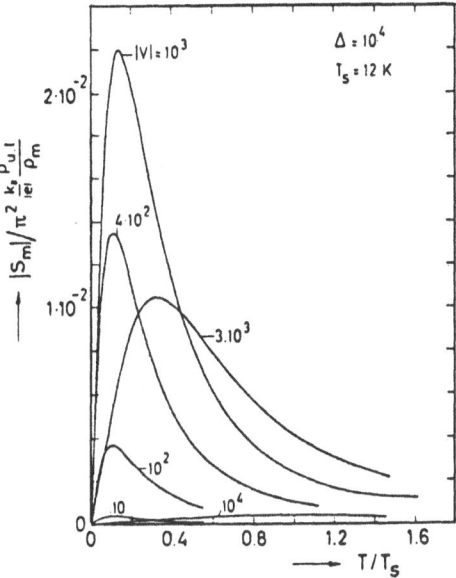

Fig. 6.13. LSF contribution to thermopower (from [74F2]). The thermopower contribution is given in terms of parameters defined in the text and in terms of ρ_{ul}, the unitarity limit for resistivity, and of ρ_m, the LSF contribution to resistivity. When the thermopower contributions are weighted by the proper resistivity terms, the net contribution from the LSF is independent of ρ_m and is a constant times the appropriate curve given in this figure.

should be noted that $|V|$ and Δ are analogous to $E_L - \eta$ and Γ, respectively. Fischer's results for the LSF contribution to thermopower are given in Figure 6.13. The peaks, in both their general form and in their general magnitudes, are similar to those which occur in Kondo alloys. When $|V| \approx \Delta$, the peak is suppressed and the contribution resembles a large term which is linear in T. This latter situation is suggestive of Zlatić and Rivier results, but additional calculations for more diverse sets of parameter values, particularly for larger values of T_S, are necessary to substantiate this suggestion.

The preceding two studies do not permit a definitive evaluation of the importance of LSF for thermoelectric behavior. However, the fact that LSF models yield the gamut of forms observed in dilute magnetic alloys indicates a better understanding of these models is essential.

6.5   Closing Comment

In this chapter we have emphasized the positive, the degree to which the behavior of magnetic alloys is understood and the qualitative and semiquantitative agreements between theory and experiment. Lest the reader be left with an erroneous impression, we conclude by quoting Suhl, who writes in the Preface of *Magnetism*, Vol. V,

The limited measure in which these theories jibe with experiment, and the considerably larger extent to which they fail to do so, becomes apparent in the chapters surveying the experimental situation.

7 | Effects of Pressure and Magnetic Field on the Thermoelectric Power

7.1 Pressure Dependence

7.1a *Introduction*

The influence of pressure on the TEP of metals is of practical and fundamental interest. The first is obvious: In a study of the effect of pressure on the properties of materials at temperatures well above or below room temperature, direct determination of the specimen temperature is requisite for meaningful data. Generally, thermocouples prove to be the most convenient

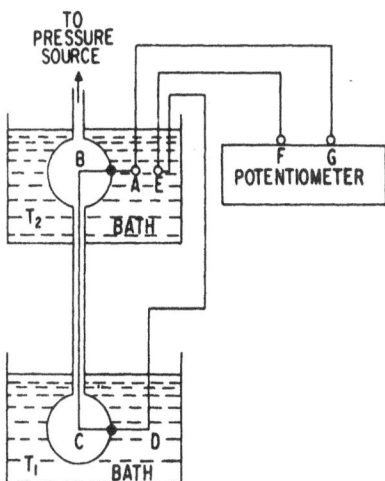

Fig. 7.1. Experimental configuration used by Bridgman (from [61B2]). *A* to *E* is a continuous wire of the material being studied.

217

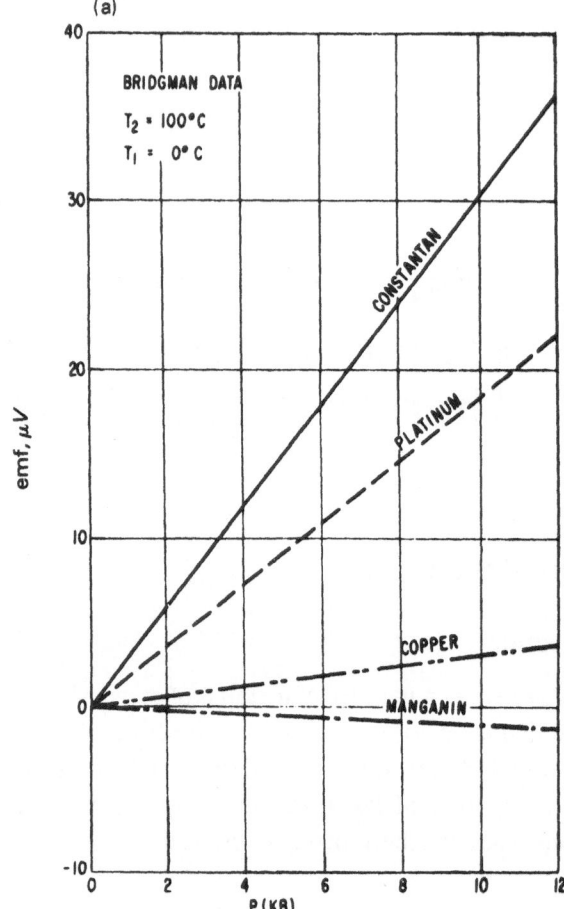

(a)

BRIDGMAN DATA

$T_2 = 100°C$

$T_1 = 0°C$

temperature sensors, and, therefore, it is important to know how high pressures might influence thermocouple calibration.

Pressure studies are also of fundamental interest because the Mott relation, Eq. (2.25), connects the TEP and the energy dependence of the conductivity, and the latter is, of course, known to depend more or less sensitively on pressure (or, to be more precise, on the volume change of the unit cell). Investigations aimed at establishing the appropriate functional relationships for pure metals and alloys have motivated these studies.

When measuring the influence of pressure on thermopower, it is essential to isolate the effects of temperature and pressure gradients. A similar problem arises in investigations of the effects of magnetic fields on the TEP as discussed in Section 7.2. One of the earliest series of experiments

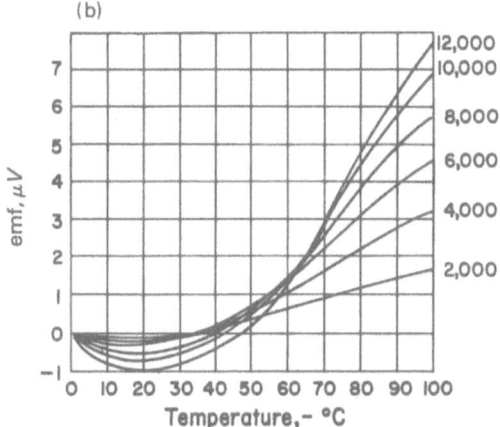

Fig. 7.2. Single wire results of Bridgman. For each material the voltage is the difference in thermal emf for a thermocouple where one leg is under pressure and the other leg is not (see Fig. 7.1). (a) The thermal emf from a constant temperature difference of 100°C is given as a function of pressure for several materials (from [61B2]). (b) The thermal emf of pure iron for a series of pressures is given as a function $T_2 - T_1$ with $T_1 \cong 0°C$ (from [49B1]). The pressures are given in kg_f/cm^2; 1 bar equals 1.0197, kg_f/cm^2.

by Bridgman [18B1, 49B1] made use of the apparatus shown in Fig. 7.1. The metal under investigation is a continuous lead between points A and E, entering the pressure vessel through seals at B and C. Copper wires, AG and EF, complete the thermoelectric circuit. Since A and B, and C and D are, pairwise, at the same temperatures T_2 and T_1, no thermoelectric emf will be generated in the portion of the circuit which has a pressure gradient. Thus the voltage developed between points A and E, measured by the potentiometer, is the difference in the thermoelectric emfs developed by the metal under study when one portion, BC, is under high pressure and the other, DE, is at atmospheric pressure. In other words, we have formed a "thermocouple" using a single substance in two distinct physical states.

7.1b Practical Thermocouples

Bridgman's work, which has withstood the test of time, was restricted to temperatures in the range 0 to 100°C, i.e., near room temperature, and was

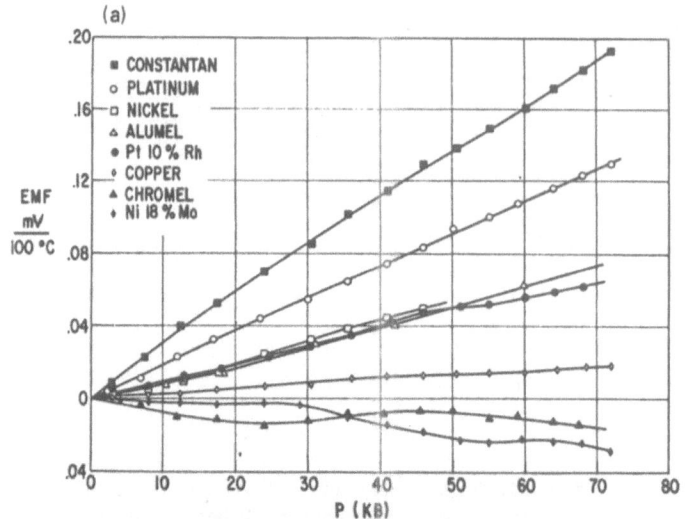

Fig. 7.3. Thermocouples under pressure (from [61B2]). (a) Changes in the absolute thermal emf for a constant temperature difference of 100°C are given as a function of pressure for some common thermocouple materials. (b) The data from (a) are combined for common thermocouple pairs and the pressure effect is converted into a temperature correction.

extended to pressures of 12 kilobars (kbar). Generally, the effects of pressure on TEP proved to be relatively minor; constantan, which displayed the largest effect, had a change in TEP of approximately $0.35 \, \mu V/K$ at 12 kbar, an average change of $0.03 \, \mu V/K$ kbar. Some of Bridgman's results are shown in Fig. 7.2. Unfortunately, most of the metals studied by Bridgman were not those commonly used for thermocouples.

Frequently, experimental circumstances involve both high pressures and high temperatures, and during the past decade a number of careful investigations have provided reliable data on various thermocouples under these conditions [65H1, 66H3, 66P1, 70G1, 71L1]. Some of the results are shown in Fig. 7.3. Evidently, as the earlier work of Bridgman had indicated, relative changes in thermopower due to pressure are fairly small. One feature is of some interest and practical concern. Lazarus [71L1], considering a number of experimental findings, concluded that chromel–alumel thermocouples suffer a substantial change above 720°C and at high pressure. This inference has direct confirmation in an earlier observation [65F3] that chromel becomes magnetic when maintained for days at high pressures near 650°C.

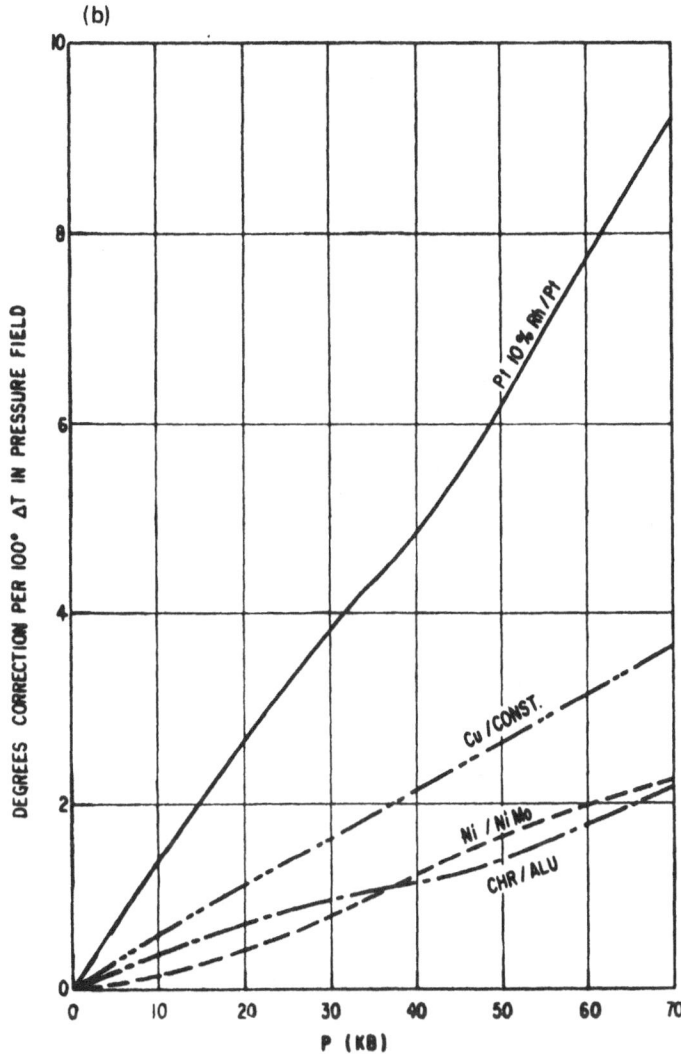

(b)

The effect of pressure on thermocouple behavior at low temperatures has received very little attention to date. Part of the reason may be that workers believed that constant-temperature environments were normally maintained in pressure vessels by various cryogenic fluids. However, an awareness of the complications that may arise from thermal gradients within the pressure vessel and also the need to measure small temperature changes when studying magnetic transitions has generated interest in such studies. At this time, only two investigations have been reported [67B7, 69B3]. The more recent of these references is more complete and contains data on copper vs. constantan and copper vs. gold–cobalt thermocouples, which show that, as in the high-temperature region, pressure effects are rather small. While these thermocouples were widely in use at one time for low-temperature work, the gold–iron vs. chromel thermocouple is now generally favored by cryogenic workers. Unfortunately no data on this couple have yet been published.*

7.1c   *Fundamental Studies*

Very few studies have been undertaken of the effect of pressure on thermopower. The work of Bridgman is one; more recently Dugdale and Mundy [61D1], Weiss and Lazarus [74W1], and Crisp *et al.* [74C1] have made such measurements on pure metals and alloys, respectively. Beyond this, the emphasis has been largely on attempts to correlate the TEP with changes of *resistivity* with pressure, a matter which has already been discussed in Chapter 2.

7.1c1   Diffusion Thermopower

Our point of departure in this section is Eq. (2.38). Using the parameter introduced by Friedel, the TEP of a pure metal is given by

$$S_d^i = \frac{\pi^2 k^2 T}{3e\eta} x_i \tag{7.1}$$

where

$$x_i = -\left(\frac{\partial \ln \rho_i}{\partial \ln \varepsilon}\right)_\eta \tag{7.2}$$

It is important to note here that these formulas have a bearing only on the diffusion contribution to the TEP. With the exception of the work of Weiss and Lazarus [74W1] to which we shall return later, all investigations to date have been performed near room temperature, and it has been tacitly assumed that S_g is negligibly small or, in any event, that the influence of pressure on this portion of the TEP may be neglected.

---

* Preliminary data at liquid nitrogen temperatures and for pressures to 4 kbar suggest that the effects are fairly small (Foiles, unpublished). However, more pronounced changes at lower temperatures, where the TEP is strongly influenced by the Kondo effect (see Chapter 6), are rather likely.

The influence of pressure arises by virtue of the change in atomic volume induced by application of high pressures. As a result, the interatomic distances are modified and, consequently, the Fermi energy in the free-electron approximation is expected to increase and the band structure may also display changes. Moreover, the phonon spectrum depends on atomic volume; in the Debye approximation this dependence is expressed through the Grüneisen constant $\gamma = -\partial \ln \theta_D / \partial \ln V$. It is convenient to separate effects due to changes in Fermi energy from those of band structure and of θ_D, at least in discussing the effects of pressure on transport, even though such a separation is not really justified.

From (7.1) we then have

$$\frac{d \ln S_d^i}{d \ln V} = -\frac{d \ln \eta}{d \ln V} + \frac{d \ln x_i}{d \ln V} \tag{7.3}$$

In the rigid-band approximation, i.e., assuming no change in band structure, it is then reasonable to proceed on the presumption that in Eq. (7.3) the second term is zero.

Setting $d \ln x_i / d \ln V = 0$, and, furthermore, assuming a free-electron model, one arrives immediately at the prediction $d \ln S_d^i / d \ln V = 2/3$.

The results of Dugdale and Mundy for the alkali metals are shown in Fig. 7.4. Initial slopes for these and the noble metals are given in Table 7.1. Clearly, the experimental results are very much at variance with expectation, and, moreover, as the data for K and Rb demonstrate, $d \ln S / d \ln V$ is not a constant.

Here again, as in the proper case of the alkali–alkali alloys (see Chapter 2), proper application of the pseudopotential approximation has led to remarkably good agreement between calculated results and experiment [67D5]. These calculations yield not only the correct signs and nearly correct magnitudes for S_d^i of the alkali metals at zero pressure but also predict the anomalously large volume dependence of the thermopower of Cs, which, according to Dickey et al., is attributed to the approach of η toward the virtual 5d-level resulting in a rapid increase of the $l = 2$ phase shift.

Except for measurements on commercial thermocouple wires, the only study of the possible dependence of the TEP of alloys on pressure is that by Crisp et al. [74C1] on some copper alloys.

7.1c2 Phonon-Drag Thermopower

Weiss and Lazarus [74W1] have measured the thermopower of sodium at pressures of 1, 2, and 3 kbar between 6 and 12 K, the temperature range

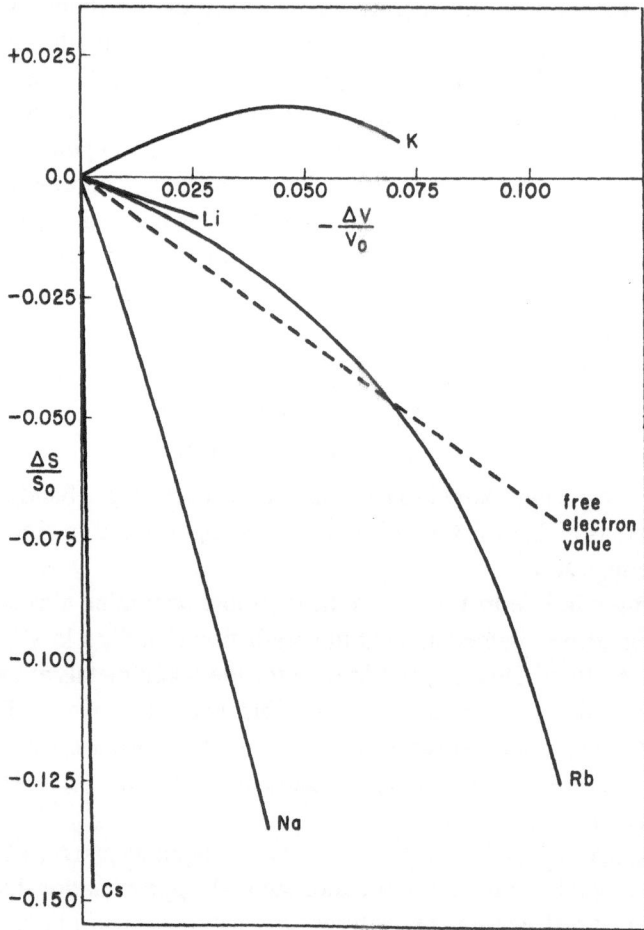

Fig. 7.4. Fractional change in thermopower vs. fractional change in volume (from [61D1]). The data are for alkali metals at 0°C.

in which S_g should be large in this metal. Although at the lower two pressures the total thermopower shows no change from the zero-pressure data, at 3 kbar S is increased slightly (i.e., less negative than at zero pressure).

In sodium, as in all metals, S_g is the sum of contributions from N and U scattering, with N processes making the larger contribution. As a sample is compressed, various factors which enter into Bailyn's theory (see Chapter 4)

Table 7.1. $d \ln S/d \ln V$ for Monovalent Metals[a]

Element	$S(0°C)$	$d \ln S/d \ln V$
Li	10.6	0.43
Na	−5.8	2.1
K	−12.9	−0.36
Rb	−9.5	0.39
Cs	−0.9	50
Cu	1.70	2.6
Ag	1.38	6.3
Au	1.79	4.4

[a] With the exception of $S(0°C)$ for the noble metals, which are taken from [58C2], these data are from [61D1].

can, in general, be expected to change, namely: (a) the speed of sound (elastic constants); (b) the phonon-polarization vectors (elastic constants); (c) electron–phonon interaction (pseudopotential); and (d) the shape of the Fermi surface.

Weiss and Lazarus argue that in sodium effects b, c, and d are negligible, and they focus their attention on the changes of the elastic constants and the concomitant change in the phonon velocity and energy. As the sound velocity increases with pressure, the phonon energies also increase and, consequently, the phonon occupation at any given temperature is reduced. Hence, both S_g^N and S_g^U are expected to be diminished in magnitude. Estimates based on the known changes in elastic constants suggest that S_g^N will decrease more rapidly than S_g^U, which is consistent with observation.

The experiments are very difficult to perform and are subject to various systematic and random errors which are discussed by Weiss and Lazarus. Moreover, sodium is not an ideal choice because it undergoes a martensitic transformation near 35 K, and data on the elastic constants of the hexagonal phase are unavailable. Finally, Weiss and Lazarus assume that between 6 and 12 K the thermopower is entirely due to phonon drag, neglecting the diffusion term completely. Although perhaps unlikely, it is possible that a change in S_d with pressure may make a significant contribution. Clearly, there is much room for further experimental work in this area.

7.2   Magnetic Field Dependence

7.2a   *Introduction*

As in the case of pressure studies, investigations of the effect of magnetic fields on thermoelectric power can be roughly divided into two categories, one concerned primarily with thermocouples of practical interest and the other with fundamental studies. The latter can be divided further into two groups: the first dealing with behavior under conditions for which quantum oscillations do not come into play and the second with studies of quantum oscillations.

7.2b   *Practical Thermocouples*

With regard to practical thermocouples, work has been concentrated largely on low-temperature couples, principally those containing gold with a fractional percentage of iron as an impurity. These thermocouples are useful at low temperatures because of the anomalously large thermopower associated with the appearance of a resistance minimum (see Section 6.3) and are now widely employed in cryogenic work [72B7, 72H2]. Unfortunately, the very effect which lends the thermocouple its high sensitivity at low temperature also makes it very susceptible to change by application of an external magnetic field.

Au + 0.07% Fe vs. chromel and constantan vs. chromel thermocouples in fields to 6 T and over the temperature range 4 to 100 K were studied by Von Middendorff [71V2]; his results are shown in Fig. 7.5. It is not clear whether his thermocouples were oriented longitudinally or transverse to the field, but, as Knittel [73K2] has shown, the field dependence of the thermopower of AuFe is not sensitive to orientation. Berman and his collaborators [63B2, 64B3] investigated the magnetic field dependence of an Au + 0.03% Fe vs. silver-normal thermocouple and of the absolute thermoelectric power of Au + 0.03% Fe for fields to 2 T and over the temperature range 1–7 K. Subsequently, Berman and Kopp [71B2] and Huntley and Walker [69H1] studied the magnetic field dependence of the thermopower of AuFe alloys with Fe concentrations between 0.5 and 1900 ppm in fields up to 7 T. This work was performed primarily to establish and evaluate impurity interaction effects in Kondo systems and has already been discussed (see Section 6.3b). Sample *et al.* [74S3] have reported on the magnetic field dependence of Au + 0.07% Fe vs. chromel P and of chromel P

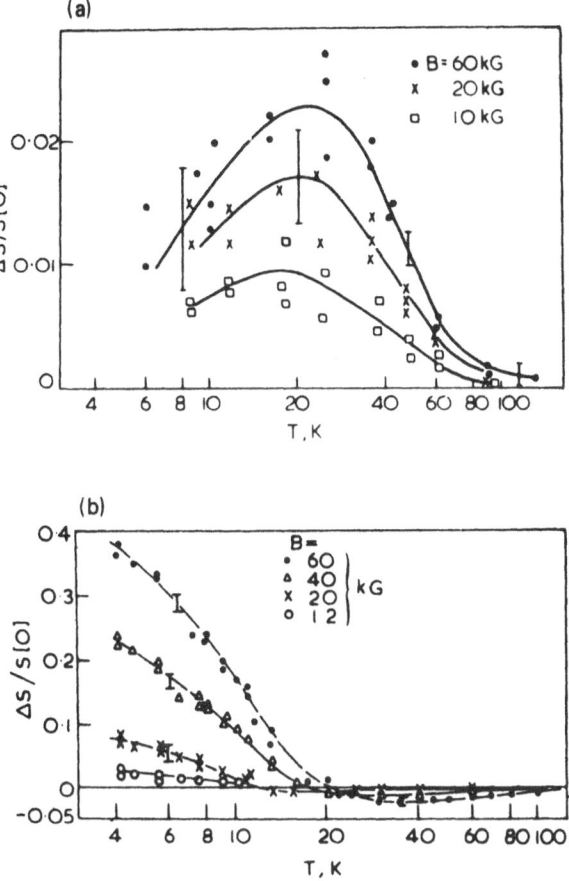

Fig. 7.5. The change of the thermoelectric power of (a) chromel–constantan and (b) gold + 0.07% Fe vs. chromel thermocouples as functions of temperature in magnetic fields to 60 kG (from [71V2]).

vs. constantan (type E) thermocouples in the temperature range 4.2–45 K and for magnetic fields up to 15 T.

Thermocouples containing AuFe arms have thermopowers that are quite sensitive to magnetic fields at temperatures below about 20 K. Sample et al., who investigated the behavior of several couples using wires taken from the same spool, also report poor reproducibility at high magnetic fields and advise against the use of the AuFe thermocouples in situations where corrections for magnetic fields must be made.

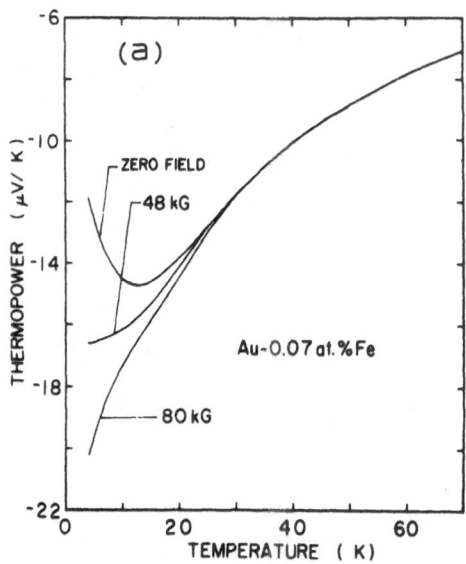

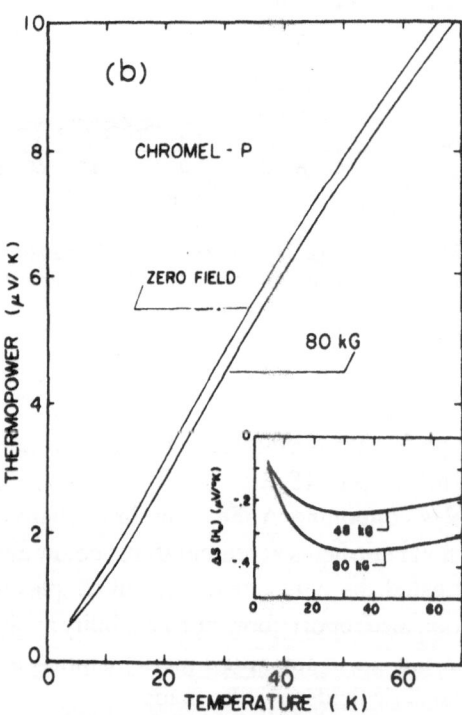

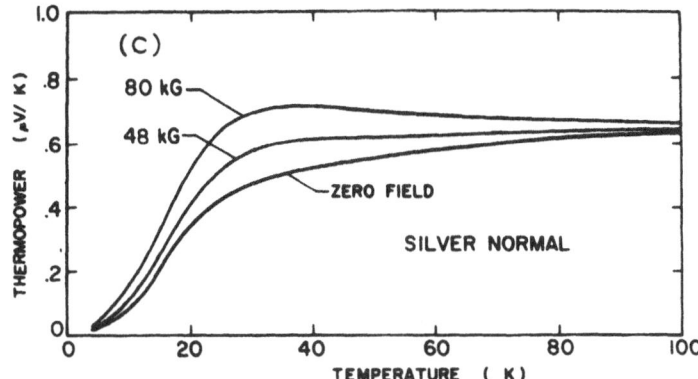

Fig. 7.6. The absolute thermoelectric power of (a) Au+0.07% Fe, (b) chromel P, and (c) silver-normal between 5 and 100 K at low, intermediate, and high magnetic fields [from 74C4].

All of the above measurements were made using the standard thermocouple configuration. Chiang, using a technique patterned after one employed by MacDonald and Pearson [57M1] in their measurement of the magnetic field dependence of the thermopower of copper at low temperatures, has succeeded in measuring the change in the absolute thermopower with field of a variety of wire samples, among them Au+0.07%Fe, silver-normal, and chromel P [74C4]. His results are in good agreement with the work of Berman *et. al.* and of Von Middendorff but show explicitly that: (a) In Au+0.07% Fe, $\Delta S(H, T)$ is large and negative at temperatures below about 20 K, reverses its sign near 25 K, and is then of a very greatly reduced magnitude at higher temperatures (Figs. 7.6a). (b) $\Delta S(H, T)$ in silver-normal is also significant and positive at all temperatures but, in contrast to Au+0.07% Fe, displays a field-dependent maximum near 25 K (Fig. 7.6c). The behavior of this material is similar to that found in pure silver, described below. (c) $\Delta S(H, T)$ of chromel P is small and negative (Fig. 7.6b), but at higher temperatures makes the dominant contribution to the change in thermo-emf of an Au+0.07% Fe vs. chromel P thermocouple with a magnetic field shown in Fig. 7.7.

To summarize, a magnetic field has relatively little influence on the thermopower of most thermocouples with the notable exception of those using AuFe alloys as one arm. The use of such thermocouples at low temperatures in environments with strong magnetic fields is, therefore, fraught with difficulties especially as the reproducibility of $\Delta S(H, T)$ is doubtful.

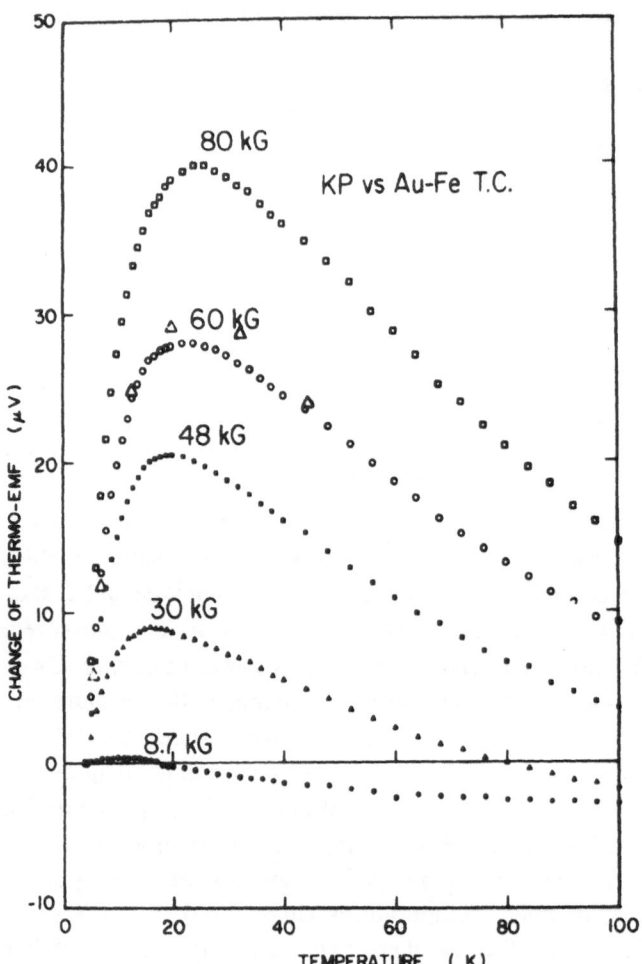

Fig. 7.7. Chromel P vs. Au + 0.07at.% Fe thermocouple. The change of thermo-emf due to transverse magnetic fields; the low temperature junction is at 4.2 K. The open triangles are the data in a longitudinal field of 60 kG from [74S3] (from [74C4]).

7.2c   Fundamental Studies

At N.R.C. in Ottawa, MacDonald and his collaborators, as part of their extensive research effort on the thermoelectric properties of monovalent metals at low temperatures, measured the change in S of sodium and of copper with magnetic field [57M1]. At the time this work was performed, type II superconductors were unknown and high-field superconducting solenoids, of course, were almost in the realm of science fiction. The limitation to fields below about 1.2 T and the then available voltage-measuring techniques severely circumscribed the accuracy of their experimental findings, and MacDonald and Pearson were forced to limit their conclusion to statements concerning the magnitude of the observed effects and their sign. In light of more recent work it is interesting to quote from their paper

the results on pure sodium and copper tempt one to suggest that the general influence of a magnetic field on a monovalent metal will be to increase the *magnitude* of the thermoelectric power, irrespective of sign.

Prompted by the observed sign change in the Hall coefficient of aluminum with magnetic field [66L1, 65C2, 69A2], Averback and Bass

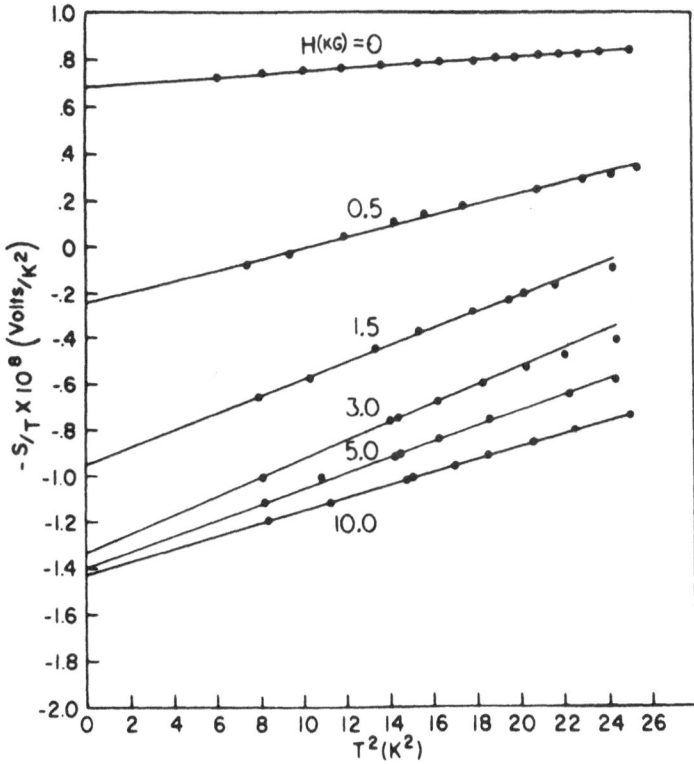

Fig. 7.8. The temperature dependence of the thermoelectric power of pure aluminum at low temperatures in transverse magnetic fields (from [73A1]).

[71A1] studied the influence of a transverse magnetic field on the thermopower of this metal. Their experimental arrangement made use of a type II superconductor as the second arm of the sample thermocouple. Thus, at temperatures below $T_c(H)$ the thermo-emf of the couple was due solely to that of aluminum, and changes with magnetic field would have to be attributed to changes in the thermopower of aluminum. Measurements ranged between about 2.5 and 6 K and showed that, as in zero field, the thermopower could be expressed as the sum of two terms, one linear and the other cubic with temperature and corresponding, presumably, to the diffusion and the phonon–drag contributions to the thermopower. Their results, shown in Fig. 7.8, revealed that both S_d and S_g depend on H. The most pronounced variation was that of S_d, which changed sign and exhibited saturation at high fields.

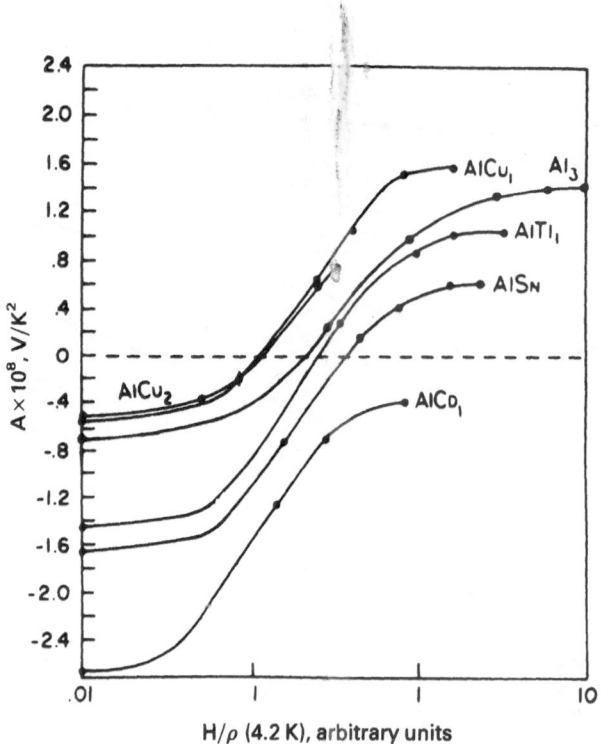

Fig. 7.9. S_d/T of aluminum and some dilute aluminum alloys vs. $H/\rho_{(4.2K)}$. The solute concentrations (in ppm) are as follows: AlCu$_1$, 340; AlCu$_2$, 50; AlTl$_1$, 100; AlSn, 100; AlCd$_1$, 25 (from [73A1]).

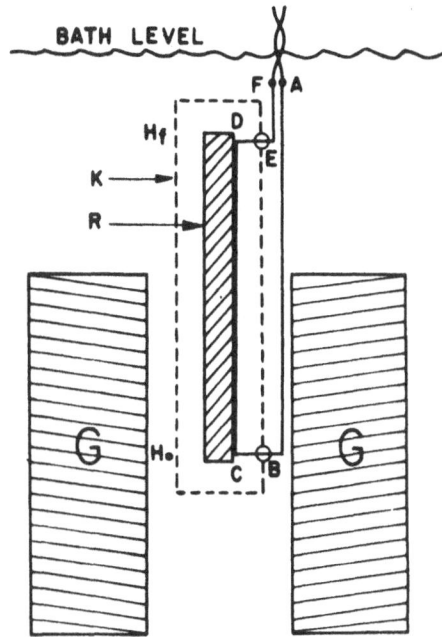

Fig. 7.10. Apparatus for measuring changes of absolute thermopower in a magnetic field. G, superconducting solenoid; K, vacuum can; R, copper rod at temperature T_0; $ABCDEF$, sample with portions: CB, at maximum field H_0 of solenoid (temperatures of C and B are T_0 and 4.2 K), DE, at fringe field H_f of solenoid (temperatures of D and E are T_0 and 4.2 K), DC, at constant temperature (from [74B3]).

Later measurements on dilute $\underline{Al}Cu$, $\underline{Al}Tl$, $\underline{Al}Sn$, and $\underline{Al}Cd$ alloys [73A1] showed that although $S_d(0, T)$—the value in zero field—was very sensitive to impurity content, the difference $\Delta S_d(H, T) = S_d(H, T) - S_d(0, T)$ as $H \rightarrow \infty$ was practically the same for all alloys (see Fig. 7.9). An explanation of this observation, based on a semiclassical treatment, was provided by Averback and Wagner [72A3]. Blatt *et al.* [74B2], using more restrictive assumptions, subsequently obtained expressions for $\Delta S_d(H, T)$ valid for all H in terms of the magnetoresistance ratio $\Delta \rho / \rho$; in the high-field limit their result reduces to that of Averback and Wagner.

Measurements of $\Delta S(H, T)$ were extended to higher temperatures by Chiang [74B3, 74C2, 74C5], who employed the experimental arrangement shown schematically in Fig. 7.10. The sample wire, about 2 m long, is brought isothermally into the field in the liquid helium bath. The wire is brought into the vacuum can through an epoxy seal at B, is wound about a cardboard former, and is thermally anchored at C to a copper rod, whose temperature can be varied by means of a heater wound near its center. Between C and D the sample wire is thermally anchored to the copper rod. It is then wound about a second former and brought out of the can through a vacuum-tight seal at E. The sample wire is attached to the leads of the

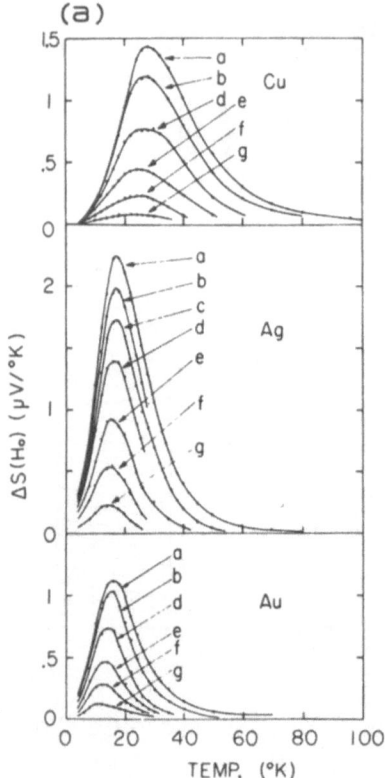
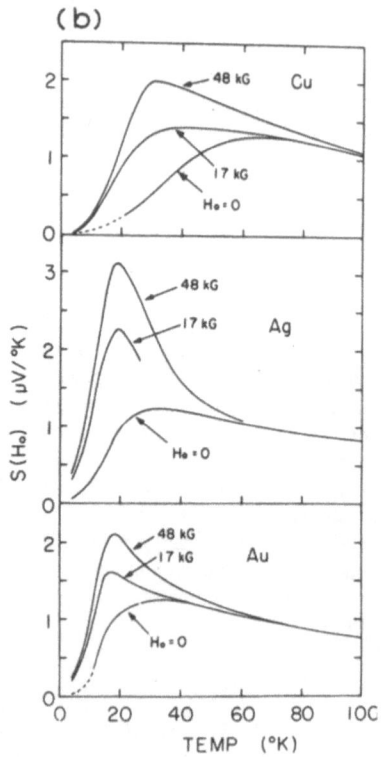

Fig. 7.11. (a) The change in absolute thermoelectric power $\Delta S(H, T)$ of copper, silver, and gold due to transverse magnetic fields of (a) 48, (b) 35, (c) 26, (d) 17.5, (e) 8.7, (f) 4.4, and (g) 1.7 kG. (b) The total absolute thermoelectric power of copper, silver and gold at 0, 17, and 48 kG (from [74B3]).

measuring system at A and F, which are junctions located adjacent to each other and in the helium bath. Since no thermal emfs can be generated in the isothermal regions, the measured emf is $V_{AF} = V_{BC} + V_{DE}$. Differentiation of V_{AF} with respect to temperature gives the difference between the thermopower in the field H_0 at the center of the magnet and in the fringing field H_f. From results of measurements at small fields H_0 one can correct for H_f and obtain

$$\Delta S(H, T) = S(H, T) - S(0, T)$$

Measurements on pure noble metals gave the surprising result that, contrary to earlier surmise, $\Delta S(H, T)$ is very large, with a maximum at a temperature which depends somewhat on field but is roughly $\theta_D/12$ for all three metals; $\Delta S(H, T)$ at a given temperature increases monotonically with H but exhibits saturation at high fields and low temperatures. The results are shown in Figs. 7.11 and 7.12. Similar results were also obtained for pure aluminum, indium, and lead [74C2, 74C5].

The temperature dependences of $S(H, T)$ and of $S(0, T)$ between 10 and 80 K strongly suggest that it is the phonon–drag term which is largely responsible for the very great change of S with magnetic field in this temperature range. Further support comes from the fact that the temperature dependence of the effect in the three noble metals scales almost perfectly with the Debye temperature.

The enhancement of S_g by a magnetic field has been explained on the basis of two alternative though somewhat related arguments. In tensor form Eq. (4.37) is

$$\sigma \cdot \mathbf{S}_g = \sum \sigma_i \cdot \mathbf{S}_g^i \qquad (7.4)$$

where $\sigma = \sum \sigma_i$ is the total electrical conductivity tensor, and σ_i and $\mathbf{S}_g^i$ are the conductivity and the phonon-drag thermopower tensors associated with the ith portion of the Fermi surface. If one now assumes that the effect of a magnetic field is to change the σ_i, in particular, in such a manner as to alter

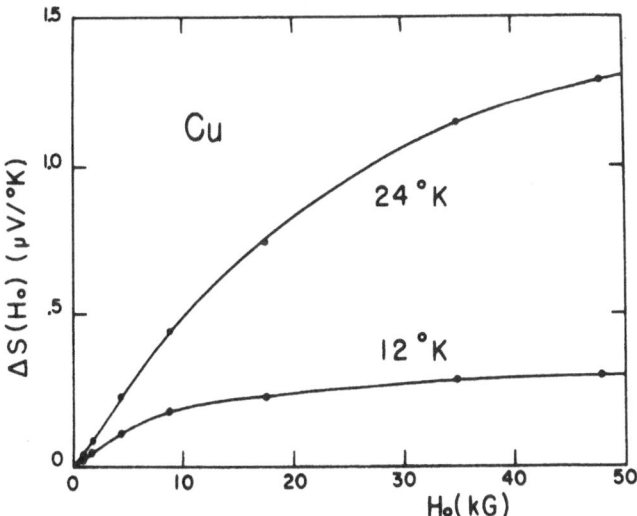

Fig. 7.12. The change in thermopower of copper with magnetic field at 12 and 24 K, showing saturation at the lower and the trend toward saturation at the higher temperature (from [74B3]).

significantly their respective contributions to σ, leaving S_g^i unaffected, physically reasonable assumptions lead to results in good agreement for Al and In [74C2]. Similarly, the results on the noble metals can be understood if we recall that the Umklapp contribution S_g^U derives largely from the neck regions, whereas S_g^N comes from the belly portions of the Fermi surface of these metals. If, as is indeed true [70K1], the electron–phonon relaxation time is shorter near the neck than the belly regions, the effect of a magnetic field, which sweeps electrons from the belly to the neck regions and vice versa [64P1], will be to enhance S_g^U. The fact that the thermopower of the noble metals increases with magnetic field is thus a natural consequence of the competition between phonon drag due to Umklapp and normal electron–phonon scattering.

Despite the strong indications that the changes of S in a magnetic field should be attributed to phonon drag, the same dilemma—is it phonon drag or phony phonon drag?—which plagues the zero-field results leads to some degree of uncertainty in the present instance. Indeed, if at zero field the usual "phonon drag" peak is in reality a diffusion contribution, the arguments of Blatt *et al.* [74B2] could account for the observations in large fields. There is some possibility that careful work on suitably selected alloys and on single crystals in moderate to high fields may help resolve this dilemma.

Whatever may be the correct explanation, it is clear that each leads to the prediction that $\Delta S(H, T)$ of a single-crystal sample should display an anisotropy which reflects that of the magnetoresistance. To date only one measurement of $\Delta S(H, T)$ on single crystals has been reported [74S4], and that work was restricted to helium temperatures where ΔS and S_g are relatively small. These results on silver clearly show the expected anisotropy of ΔS and appear to be in qualitative agreement with the higher-temperature work on polycrystalline wires.

7.2d   Landau Quantization Effects

As is well known, when $\omega\tau > 1$ and $kT < \beta H$, where $\beta = eh/m^*c$, all electronic properties of a metal display oscillatory properties which reflect the quantization of electron orbits in the magnetic field [68G1, 69S4, 69P2]. Since β and ω in a real metal depend on the orientation of H relative to the crystallographic axes, quantum effects manifest themselves only in experiments on single crystals; among the better known are the de Haas–van Alphen effect [68G1] and the Shubnikov–de Haas effect [35D1, 60P2].

Table 7.2. Field Dependence of the Adiabatic and Isothermal Diffusion Thermoelectric Power Tensors (H is along the z-direction)

Fermi surface topology	Diffusion thermopower tensor					
	Isothermal			Adiabatic		
Closed orbits compensated metals	H^1	H^1	H^1	H^3+H^2	H^3+H^2	H^2+H^1
	H^1	H^1	H^1	H^3+H^2	H^3+H^2	H^2+H^1
	H^0	H^0	H^0	H^2+H^1	H^2+H^1	H^1
Closed orbits uncompensated metals	H^0	H^{-1}	H^0	H^0	H^1	H^0
	H^{-1}	H^0	H^0	H^1	H^0	H^0
	H^{-1}	H^{-1}	H^0	H^0	H^0	H^0
Open orbits along x-direction	H^0	H^1	H^1	H^2	H^1	H^1
	H^{-1}	H^0	H^0	H^1	H^0	H^0
	H^{-1}	H^0	H^0	H^1	H^0	H^0

In addition to the oscillatory effects, the transport properties of metal single crystals also display, as a rule, very pronounced anisotropies which are intimately related to the topological features of the Fermi surface. Over the past decades a plethora of investigations has provided a fairly complete and detailed picture of the Fermi surface of nearly all pure metals.

As might be expected, not only the resistance and the Hall voltage but also the thermoelectric effects exhibit both quantum oscillations and anisotropies related to the Fermi surface topology. Bychkov, Gurevich, and Nedlin [60B3] first discussed the effects of open and closed orbits on the thermoelectric power tensor, and their conclusions were further extended by Woollam [69W2], who provided results for the field dependence of the isothermal and the adiabatic thermoelectric power tensor, which are summarized in Table 7.2.

Measurements by Woollam [69W2] of the thermoelectric and the Nernst–Ettinghausen voltages on single-crystal tin, a compensated metal with open orbits along certain crystallographic directions, showed the expected anisotropy and confirmed the predicted H^3+H^2 dependence of the adiabatic Nernst–Ettinghausen coefficient.

With regard to the quantum oscillations, thermoelectric properties have some very distinct advantages over more conventional transport effects. In brief, oscillatory electronic properties arise because, as H is varied, the quantized Landau levels repeatedly pass through the Fermi energy, resulting in a variation (periodic in $1/H$) of the density of states at the Fermi energy. Since specific heat, magnetic susceptibility, resistivity, etc., all depend on $N(\eta)$, such a periodic variation is reflected, therefore, in these physical properties. A detailed theory of quantum oscillations of the thermoelectric effects has been given by Zil'berman [56Z1] and in an unpublished thesis by Horton [64H2]. The unique aspect of thermoelectric parameters in relation to oscillatory behavior is that they depend on the *derivative* of $N(\varepsilon)$ evaluated at the Fermi energy. Consequently, whereas Shubnikov–de Haas oscillations are superimposed on a large background magnetoresistance, Landau quantization causes the thermopower to oscillate between positive and negative extrema, i.e., the amplitude of the oscillations is substantially larger than the average thermopower; the same is true for the Peltier coefficient as well.

The semimetal bismuth has traditionally been the substance in which oscillatory effects were first studied, and thermopower oscillations were also first observed in bismuth [55S2, 63G4]. More recently, oscillations have been studied in antimony [65L2], in tin [69W2], and in aluminum [74K1], and Peltier oscillations have been observed in zinc [69T2].

Note Added in Proof

One consequence of recent work on magnetothermoelectric effects deserves mention here because of its fundamental importance to general transport theory.

During the late fifties and early sixties much theoretical activity concentrated on the question of electron mass enhancement due to electron–phonon interaction. Very crudely, phonon induced mass enhancement may be viewed as follows: An electron, by virtue of its coulomb field, attracts nearby positive ions, i.e., induces a polarization cloud; as the electron moves through the lattice, this polarization cloud must move along with it. The effect is equivalent to an increase in the electron's effective mass.

This phonon induced enhancement results in an increase in the density of states $N(\eta)$ [see Eq. (2.4)] and, hence, influences the electronic specific heat and other properties of the electron gas. Using a heuristic approach, Prange and Kadanoff [64P2] concluded that electron–phonon mass enhancement will, however, not be reflected in d.c. transport properties. Holstein [64H3] was able to prove this rigorously for the electrical conductivity, but such rigorous proofs were not developed for the thermal conductivity or thermoelectric power.

As shown in Fig. 7.9, for dilute polycrystalline alloys of aluminum between 1 and 5 K, $\Delta S_d = S_d(H \rightarrow \infty) - S_d(H = 0)$ is nearly independent of impurity even though $S_d(H \rightarrow \infty)$ and $S_d(H = 0)$ do, separately, depend sensitively on impurity content. A qualitative explanation for this was given by Averback and Wagner [72A3], but their predicted value of ΔS_d was roughly 50% below the experimental results.

It has now been demonstrated by Opsal *et al.* [76O1] that this discrepancy is not due to the use of polycrystalline samples or to computational approximations employed by Averback and Wagner. Instead, Opsal was able to give sound heuristic argument to show that although in the calculation of electrical and thermal conductivities mass enhancement corrections do, indeed, cancel out, such cancellation does *not* occur in the expression for the thermopower $S_d(H)$. Inclusion of mass enhancement then increases the theoretical prediction for ΔS_d by about 50% and brings theory and experiment into excellent accord. Stimulated by this work, Holstein and Lyo [76H1] have reexamined the microscopic theory and obtained results compatible with those of Opsal *et al.*

It now appears to be conclusively established that, contrary to previous belief, electron mass renormalization due to electron–phonon interaction does manifest itself in thermoelectricity and, therefore, may be studied by measurements of this phenomenon. Moreover, previous calculations of thermopower which have neglected this mass enhancement should be corrected to include this effect.

References

'26S1 A. Seebeck (1826), *Pogg. Ann.* **6**, 133.

'34P1 J. C. Peltier (1834), *Ann. Chim. Phys.* **56**, 371.

'82K1 Lord Kelvin, *Collected Papers I* Cambridge University Press, (1882).

18B1 P. W. Bridgman (1918), *Proc. Amer. Acad. Arts Sci.* **53**, 269.

28B1 G. Borelius, W. H. Keesom, and C. H. Johansson (1928), *Proc. Acad. Sci. Amst.* **31**, 1046.

30B1 G. Borelius, W. H. Keesom, C. H. Johansson, and J. O. Linde (1930), *Proc. Acad. Sci. Amst.* **33**, 17.

31B1 G. Borelius, W. H. Keesom, C. H. Johansson, and J. O. Linde (1931), *Proc. Acad. Sci. Amst.* **34**, 1365.

31O1 L. Onsager (1931), *Phys. Rev.* **37**, 405; **38**, 2265.

32B1 G. Borelius, W. H. Keesom, C. H. Johansson, and J. O. Linde (1932), *Proc. Acad. Sci. Amst.* **35**, 10.

32G1 K. E. Grew (1932), *Phys. Rev.* **41**, 356.

33D1 W. J. de Haas, J. H. de Boer, and G. J. van den Berg (1933), *Physica* **1**, 1115.

34B1 P. W. Bridgeman (1934), *The Thermodynamics of Electrical Phenomena in Metals*, Macmillan Co.

35D1 W. J. de Haas, J. W. Blom, and L. Shubnikov (1935), *Physica* **2**, 907.

35N1 L. Nordheim and C. J. Gorter (1935), *Physica* **2**, 383.

36M1 N. F. Mott (1936), *Proc. R. Soc. Lond. A* **156**, 368.

37B1 W. G. Baber (1937), *Proc. R. Soc. Lond. A* **158**, 383.

38S1 E. C. Stoner (1938), *Proc. R. Soc. Lond. A* **165**, 372.

41P1 H. H. Potter (1941), *Proc. Phys. Soc.* **53**, 695.

47C1 H. B. G. Casimir and A. Rademakers (1947), *Physica* **13**, 133.

47N1 J. Nyström (1947), *Ark. for Mat. Astr. och Fys.* **34A**, #27, 1.

48C1 H. B. Callen, *Phys. Rev.* **73**, 1349.

48L1 J. J. Lander (1948), *Phys. Rev.* **74**, 479.

49B1 P. W. Bridgman (1949), *The Physics of High Pressure*, G. Bell and Sons, Ltd., London, England.

49K1 M. Kohler (1949), *Z. Phys.* **126**, 481.

51W1 H. W. Worner (1951), *Austra. J. Sci. Res. A* **4**, 62.

52A1 H. K. Adenstedt (1952), *Trans. Am. Soc. Met.* **44**, 949.

52C1 H. B. Callen (1952), *Phys. Rev.* **85**, 16.

52P1 A. B. Pippard and G. T. Pullan (1952), *Proc. Camb. Philos. Soc.* **48**, 188.

53D1 C. A. Domenicali (1953), *Phys. Rev.* **92**, 877.

53F1 J. Friedel (1953), *J. Phys. Radium* **14**, 561.

53M1 D. K. C. MacDonald and W. B. Pearson, (1953), *Proc. R. Soc. Lond. A* **219**, 373.

53W1 A. H. Wilson (1953), *The Theory of Metals*, Cambridge University Press, Cambridge.

54M1 D. K. C. MacDonald and W. B. Pearson, (1954), *Philos. Mag.* **45**, 491.

55B1 M. Blackman (1955), *Encyclopedia of Physics*, Vol. VII, part 1, Springer Verlag, Berlin.

55K1 P. G. Klemens (1955), *Proc. Phys. Soc. Lond. A* **68**, 1113.

55S1 H. Shenker, J. Z. Lauritzen, Jr., R. J. Coruccini, and S. T. Longerber (1955), *Reference Tables for Thermocouples*, Nat. Bur. Stand. Circ. 561.

55S2 M. C. Steele and J. Babiskin (1955), *Phys. Rev.* **98**, 359.

55T1 I. M. Templeton (1955), *J. Sci. Instrum.* **32**, 314.

56D1 P. de Faget, de Casteljau, and J. Friedel (1956), *J. Phys. Radium* **17**, 27.

56F1 J. Friedel (1956), *Can. J. Phys.* **34**, 1190.

56K1 T. Kasuya (1956), *Prog. Theor. Phys.* **16**, 58.

56K2 P. G. Klemens (1956), *Encyclopedia of Physics*, Vol. XIV, 198, Springer Verlag, Berlin.

56O1 F. A. Otter (1956), *J. Appl. Phys.* **27**, 197.

56S1 E. H. Sondheimer (1956), *Can. J. Phys.* **34**, 1246.

56T1 J. C. Taylor and B. R. Coles, (1956), *Phys. Rev.* **102**, 27.

56Z1 G. E. Zil'berman (1956), *Zh. Eksp. Tem. Fiz. [Sov. Phys. JETP]* **2**, 650.

57D1 A. R. De Vroomen and C. Van Baarle (1957), *Physica* **23**, 785.

57M1 D. K. C. MacDonald and W. B. Pearson, (1957), *Proc. R. Soc. Lond. A* **241**, 257.

57M2 N. F. Mott and K. W. H. Stevens (1957), *Philos. Mag.* **2**, 1364.

58B1 M. Bailyn (1958), *Phys. Rev.* **112**, 1587.

58C1 J. W. Christian, J. P. Jan, W. B. Pearson, and I. M. Templeton, (1958), *Proc. R. Soc. Lond. A* **245**, 213.

58C2 N. Cusack and P. Kendall (1958), *Proc. Phys. Soc.* **72**, 898.

58C3 B. R. Coles (1958), *Adv. Phys.* **7**, 40.

58D1 C. A. Domenicali (1958), *Phys. Rev.* **112**, 1863.

58J1 J. P. Jan, W. B. Pearson, and I. M. Templeton (1958), *Can. J. Phys.* **36**, 627.

58K1 P. G. Klemens in *Solid State Physics*, Vol **7**, 1, edited by Seitz and Turnbull, Academic Press, New York, 1958.

58M1 D. K. C. MacDonald, W. B. Pearson, and I. M. Templeton (1958), *Proc. Roy. Soc. Lond. A* **248**, 107.

58M2 D. K. C. MacDonald, W. B. Pearson, and I. M. Templeton (1958), *Philos. Mag.* **3**, 657.

58M3 D. K. C. MacDonald, W. B. Pearson and I. M. Templeton (1958), *Philos. Mag.* **3**, 917.

58M4 A. B. Migdal (1958), *Zh. Eksp. Teor. Fiz. [Sov. Phys.-JETP]* **7**, 996.

58P1 W. B. Pearson (1958), *A Handbook of Lattice Spacings and Structures of Metals and Alloys*, Pergamon Press, London.

59J1 H. A. Johansen and R. C. Miller (1959), *J. Less Common Met.* **1**, 331.

59K1 T. Kasuya (1959), *Prog. Theor. Phys.* **22**, 227.

60B1 M. Bailyn (1960), *Philos. Mag.* **5**, 1059.

60B2 F. J. Blatt and R. H. Kropschot (1960), *Phys. Rev.* **118**, 480.

60B3 Yu A. Bychkov, L. E. Gurevitch, and G. M. Nedlin (1960), *Zh. Eksp. Teor. Fiz. [Sov. Phys.-JETP]* **10**, 377.

60D1 De Vroomen, C. Van Baarle, and A. J. Cuelenaere (1960), *Physica* **26**, 19.

60G1 A. V. Gold, D. K. C. MacDonald, W. B. Pearson, and I. M. Templeton (1960), *Philos. Mag.* **5**, 765.

60M1 D. K. C. MacDonald, W. B. Pearson, and I. M. Templeton (1960), *Proc. R. Soc. Lond. A* **256**, 334.

60P1 W. B. Pearson (1960), *Can. J. Phys.* **38**, 1048.

60P2 A. B. Pippard (1960), *Rep. Prog. Phys.* **23**, 176.

60Z1 J. M. Ziman (1960), *Electrons and Phonons*, Oxford University Press, England.

61A1 P. W. Anderson (1961), *Phys. Rev.* **124**, 41.

61B1 H. J. Born, S. Legvold, and F. H. Spedding (1961), *J. Appl. Phys.* **32**, 2543.

61B2 F. P. Bundy (1961), *J. Appl. Phys.* **32**, 483.

61C1 P. Carruthers (1961), *Rev. Mod. Phys.* **33**, 92.

61D1 J. S. Dugdale and J. N. Mundy (1961), *Philos. Mag.* **6**, 1463.

61D2 A. R. De Vroomen and M. L. Potters (1961), *Physica* **27**, 1083.

61G1 A. M. Guénault and D. K. C. MacDonald (1961), *Proc. R. Soc. Lond. A* **264**, 41.

61G2 A. M. Guénault and D. K. C. MacDonald (1961), *Philos. Mag.* **6**, 1201.

61G3 A. V. Gold and W. B. Pearson (1961), *Can. J. Phys.* **39**, 445.

61P1 R. L. Powell, M. D. Bunch, and R. J. Corruccini (1961), *Cryogenics* **1**, 139.

61S1 B. Segall (1961), *Phys. Rev.* **125**, 109.

62B1 M. Bailyn (1962), *Phys. Rev.* **126**, 2040.

62K1 N. V. Kolomoets and M. V. Vedernikov (1962), *Fiz. Tverd.* [*Sov. Phys.-Solid State*] **3**, 1996.

62L1 J. C. Lachman and J. A. McGurty (1962), *Temperature, Its Measurement and Control in Science and Industry*, Vol. III, 2, 177, Reinhold Publishing Corp. New York.

62M1 D. K. C. MacDonald (1962), *Thermoelectricity: An Introduction to the Principles* John Wiley and Sons, New York.

62M2 D. K. C. MacDonald, W. B. Pearson, and I. M. Templeton (1962), *Proc. R. Soc. Lond. A* **266**, 161.

62R1 D. J. Roaf (1962), *Philos. Trans. R. Soc. Lond.* **255**, 135.

63B1 G. A. Burdick (1963), *Phys. Rev.* **129**, 138.

63B2 R. Berman and D. J. Huntley (1963), *Cryogenics* **3**, 70.

63C1 E. L. Christenson (1963), *J. Appl. Phys.* **34**, 1485.

63G1 A. M. Guénault and D. K. C. MacDonald (1963), *Proc. R. Soc. Lond. A* **274**, 154.

63G2 D. A. Goodings (1963), *Phys. Rev.* **132**, 542.

63G3 D. Griffiths and B. R. Coles (1963), *Proc. Phys. Soc.* **82**, 127.

63G4 C. G. Grenier, J. M. Reynolds, and J. R. Sybert (1963), *Phys. Rev.* **132**, 58.

63H1 W. G. Henry and P. A. Schroeder (1963), *Can. J. Phys.* **41**, 1076.

63M1 A. R. Mackintosh and L. R. Sill (1963), *J. Phys. Chem. Solids* **24**, 501.

64B1 F. J. Blatt (1964), *Proc. Phys. Soc.* **83**, 1065.

64B2 F. J. Blatt, M. Garber, and B. W. Scott (1964), *Phys. Rev. A* **136**, 729.

64B3 R. Berman, J. C. F. Brock, and D. J. Huntley (1964), *Cryogenics* **4**, 233.

64C1 R. S. Crisp, W. G. Henry, and P. A. Schroeder (1964), *Philos. Mag.* **10**, 553.

64G1 L. E. Gurevich and I. Y. Korenblit (1964), *Fiz. Tverd.* [*Sov. Phys.-Solid State*] **6**, 1960.

64H1 R. P. Huebener (1964), *Phys. Rev. A* **135**, 1281.

64H2 P. B. Horton (1964), Ph.D. Dissertation, Louisiana State University.

64H3 T. Holstein (1964), *Ann. Phys.* **29**, 410.

64K1 W. Koster and H. P. Rave, (1964), *Z. Metallkd.* **55**, 750.

64K2 J. Kondo, (1964), *Prog. Theor. Phys.* **32**, 37.

64M1 N. F. Mott (1964), *Adv. Phys.* **13**, 325.

64N1 O. P. Naumkin, V. F. Terekhova, and Ye. M. Savitsky, (1964), *Fiz. Met. Metalloved.* [*Phys. Met. Metallogr*] **16**, 22.

64P1 A. B. Pippard (1964), *Proc. R. Soc. Lond. A* **202**, 464.

64P2 R. E. Prange and L. P. Kadanoff (1964), *Phys. Rev.* **134**, A566.

64R1 M. Roesler (1964), *Phys. Status solidi* **7**, K75.

64V1 C. Van Baarle, A. J. Cuelenaere, C. J. Roest, and M. K. Young (1964), *Physica* **30**, 244.

64W1 J. D. Wasscher and C. Haas (1964), *Phys. Lett.* **8**, 302.

64Z1 J. E. Zimmermann and A. H. Silver (1964), *Phys. Lett.* **10**, 47.
65C1 R. S. Crisp and W. G. Henry (1965), *Philos. Mag.* **11**, 841.
65C2 J. N. Cooper, P. Cotti, and F. B. Rasmussen (1965), *Phys. Lett.* **19**, 560.
65D1 M. Dixon, F. E. Hoare, T. M. Holden, and D. E. Moody (1965), *Proc. R. Soc. Lond. A* **285**, 561.
65D2 J. S. Dugdale (1965), *Physics of Solids at High Pressures,* edited by C. T. Tomizuka and R. M. Emrick, Academic Press, Inc., New York.
65D3 E. Daniel and J. Friedel (1965), *Proc. L.T.9,* Plenum Press, New York, **B**, 933.
65F1 R. Fletcher and D. Greig (1965), *Phys. Lett.* **17**, 6.
65F2 R. E. Fryer, C. C. Lee, V. Rowe, and P. A. Schroeder (1965), *Physica* **31**, 1491.
65F3 C. L. Foiles and C. T. Tomizuka (1965), *J. Appl. Phys.* **36**, 3839.
65G1 D. Greig and J. P. Harrison (1965), *Proc. L.T.9,* Plenum Press, New York, **B**, 1050.
65H1 R. E. Hanneman and H. M. Strong (1965), *J. Appl. Phys.* **36**, 523.
65K1 J. Kondo (1965), *Prog. Theor. Phys.* **34**, 372.
65L1 C. C. Lee (1965), M.S. Dissertation, Michigan State University.
65L2 J. R. Long, C. G. Grenier, and J. M. Reynolds (1965), *Phys. Rev. A* **140**, 187.
65M1 G..T. Meaden (1965), *Electrical Resistance of Metals,* Plenum Press, New York.
65N1 Y. Nagaoka (1965), *Phys. Rev. A* **138**, 1112.
65R1 P. Radhakrishna and M. Nielsen (1965), *Phys. Status Solidi* **11**, 111.
65S1 P. A. Schroeder, R. Wolf, and J. Á. Woollam (1965), *Phys. Rev.* **138**, A105.
65S2 L. R. Sill and S. Legvold (1965), *Phys. Rev.* **137**, A1139.
65W1 I. Weinberg and C. W. Shultz (1965), *J. Phys. Chem. Solids* **27**, 474.
65W2 W. Worobey, P. Lindenfeld, and B. Serin (1965), *Phys. Lett.* **16**, 15.
65W3 J. Weinberg (1965), *Phys. Rev. A* **139**, 838.
66B1 N. F. Berk and J. R. Schrieffer (1966), *Phys. Rev. Lett.* **17**, 433.
66B2 G. Boato, M. Bugo, and C. Rizzuto (1966), *Nuovo Cimento,* **X45**, 226.
66C1 J. Clarke (1966), *Philos. Mag.* **13**, 115.
66D1 S. Doniach and S. Engelsberg (1966), *Phys. Rev. Lett.* **17**, 750.
66D2 J. S. Dugdale and A. M. Guénault (1966), *Philos. Mag.* **13**, 503.
66G1 E. S. R. Gopal (1966), *Specific Heats at Low Temperatures,* Heywood Books, London.
66G2 A. A. Gomes (1966), *J. Phys. Chem. Solids* **27**, 451.
66H1 C. Herring (1966), *Magnetism,* Vol. IV, 298, Academic Press, New York.
66H2 R. P. Huebener (1966), *Phys. Rev.* **146**, 490.
66H3 R. E. Hanneman and H. M. Strong (1966), *J. Appl. Phys.* **37**, 612.
66K1 N. V. Kolomoets (1966), *Fiz. Tverd. [Sov. Phys. Solid State]* **8**, 799.
66L1 R. Luck (1966), *Phys. Status Solidi* **18**, 49.
66M1 D. L. Mills and P. Lederer (1966), *J. Phys. Chem. Solids* **27**, 1805.
66P1 E. T. Peters and J. J. Ryan (1966), *J. Appl. Phys.* **37**, 933.
66V1 C. Van Baarle, C. J. Roest, M. K. Roest-Young, and F. W. Gorter (1966), *Physica* **32**, 1700.
66Z1 J. E. Zimmermann and A. H. Silver (1966), *Phys. Rev.* **141**, 367.
67A1 A. T. Aldred (1967), *J. Phys. Soc. Japan* **22**, 762.
67B1 M. Bailyn (1967), *Phys. Rev.* **157**, 480.
67B2 F. J. Blatt and W. H. Lucke (1967), *Philos. Mag.* **15**, 649.
67B3 F. J. Blatt, M. Garber, and B. Scott (1967), *Progress in Engineering Science,* Gordon and Breach, New York, 683.
67B4 F. J. Blatt, D. J. Flood, V. Rowe, P. A. Schroeder, and J. E. Cox (1967), *Phys. Rev. Lett.* **18**, 395.
67B5 G. Boato and J. Vig (1967), *Solid State Commun.* **5**, 649.
67B6 C. M. Bhandari and G. S. Verma (1967), *Nuovo Cimento B* **47**, 129.

67B7 D. Block and F. Chaisse (1967), *J. Appl. Phys.* **38**, 409.

67C1 I. A. Campbell, A. Fert, and A. R. Pomeroy (1967), *Philos. Mag.* **15**, 977.

67C2 P. P. Craig, W. I. Goldberg, T. A. Kitchens, and J. I. Budnick (1967), *Phys. Rev. Lett.* **19**, 1334.

67C3 J. E. Cox and W. H. Lucke (1967), *J. Appl. Phys.* **38**, 3851.

67D1 J. S. Dugdale and M. Bailyn (1967), *Phys. Rev.* **157**, 485.

67D2 J. S. Dugdale and Z. S. Basinski (1967), *Phys Rev.* **157**, 552.

67D3 M. D. Daybell and W. A. Steyert (1967), *Phys. Rev. Lett.* **18**, 398.

67D4 M. D. Daybell, D. L. Kohlstedt, and W. A. Steyert (1967), *Solid State Commun.* **5**, 871.

67D5 J. M. Dickey, A. Meyer, and W. H. Young (1967), *Proc. Phys. Soc.* **92**, 460.

67F1 M. E. Fisher (1967), *Rep. Prog. Phys.* **30**, 615.

67F2 C. L. Foiles (1967), *Rev. Sci. Instrum.* **38**, 731.

67F3 K. Fischer (1967), *Phys. Rev.* **158**, 613.

67G1 A. Guénault (1967), *Philos. Mag.* **15**, 17.

67G2 R. J. Gripshover, J. B. Van Zytveld, and J. Bass (1967), *Phys. Rev.* **163**, 598.

67G3 D. Gainon, P. Donze, and J. Sierro (1967), *Solid State Commun.* **5**, 151.

67H1 P. Handler, D. E. Mapother, and M. Rayl (1967), *Phys. Rev. Lett.* **19**, 356.

67M1 A. Meyer, C. W. Nestor, and W. H. Young (1967), *Proc. Phys. Soc.* **92**, 446.

67N1 Y. Nagaoka (1967), *Prog. Theor. Phys.* **37**, 13.

67R1 J. E. Robinson (1967), *Phys. Rev.* **161**, 533.

67R2 C. F. Ratto and A. Blandin (1967), *Phys. Rev.* **156**, 513.

67S1 A. I. Schindler and M. J. Rice (1967), *Phys. Rev.* **164**, 759.

67S2 K. Schröder and M. Yessik (1967), *J. Phys. Chem. Solids* **28**, 1713.

67S3 B. W. Scott, F. J. Blatt, and M. Garber (1967), *Recent Advances in Engineering Science*, Gordon and Breach, New York.

67S4 H. Suhl and D. Wong (1967), *Physics* **3**, 17.

67S5 H. Suhl (1967), *Rendiconti de la Scuola Interazionale di Fisiça*, Academic Press, London.

67V1 C. Van Baarle, F. W. Gorter, and P. Winsenius (1967), *Physica* **35**, 223.

67V2 C. Van Baarle (1967), *Physica* **33**, 424.

67W1 I. Weinberg (1967), *Phys. Rev.* **157**, 564.

68B1 F. J. Blatt (1968), *Physics of Electronic Conduction in Solids*, McGraw-Hill Book Co., New York.

68C1 A. D. Caplin, C. L. Foiles, and J. Penfold (1968), *J. Appl. Phys.* **39**, 842.

68F1 T. Farrell and D. Greig (1968), *J. Phys. C* **1**, 1359.

68F2 M. E. Fisher and J. S. Langer (1968), *Phys. Rev. Lett.* **20**, 665.

68F3 R. Fletcher and D. Greig (1968), *Philos. Mag.* **17**, 21.

68F4 C. L. Foiles and A. I. Schindler (1968), *Phys. Lett. A* **26**, 154.

68F5 T. Farrell, D. Greig, and J. A. Rowlands (1968), *Proc. LT11*, St. Andrews University Press, **2**, 1074.

68F6 L. F. Feigenbutz (1968), *J. Phys. E* **1**, 489.

68G1 A. V. Gold (1968), *Solid State Physics*, edited by J. F. Cochran and R. R. Haering, Gordon and Breach, Inc., New York, 39.

68G2 J. C. Garland and R. Bowers (1968), *Phys. Rev. Lett.* **21**, 1007.

68H1 A. Hasegawa and T. Kasuya (1968), *J. Phys. Soc. Japan* **25**, 141.

68H2 R. P. Huebener (1968), *Phys. Rev.* **171**, 634.

68K1 Yu. Kagan, A. P. Zheinov, and H. Paskaev (1968), in *Localized Excitations in Solids*, edited by R. F. Wallis, Plenum Press, New York, 675.

68L1 P. Lederer and D. L. Mills (1968), *Phys. Rev.* **165**, 837.

68L2 P. Leonard (1968), *Phys. Lett. A* **27**, 641.

68N1 P. E. Nielsen and P. L. Taylor (1968), *Phys. Rev. Lett.* **21**, 893.

68N2 H. Nagasawa (1968), *J. Phys. Soc. Japan* **25**, 691.

68R1 J. E. Robinson and J. D. Dow (1968), *Phys. Rev.* **171**, 815.

68S1 K. Schröder and H. Tomaschke (1968), *Phys. Kondens. Mater.* **7**, 318.

68S2 L. L. Sparks, R. L. Powell, and W. J. Hall (1968), *U.S. Nat. Bur. Stand. Report* #9712.

68S3 L. L. Sparks and W. J. Hall (1968), *U.S. Nat. Bur. Stand. Report* #9719.

68T1 D. E. Thornton, W. H. Young, and A. Meyer (1968), *Phys. Rev.* **166**, 746.

68T2 A. L. Trego and A. R. Mackintosh (1968), *Phys. Rev.* **166**, 495.

68W1 L. S. Wright (1968), *Can. J. Phys.* **46**, 1711.

68Z1 D. A. Zych (1968), *Rev. Sci. Instrum.* **39**, 1968.

69A1 R. Aoki and T. Ohtsuka (1969), *J. Phys. Soc. Japan* **26**, 651.

69A2 N. W. Ashcroft (1969), *Phys. Kondens. Mater.* **9**, 45.

69B1 C. M. Bhandari and G. S. Verma (1969), *Nuovo Cimento B* **60**, 249.

69B2 E. Brewig, W. Kierspe, V. Schotte, and O. Wagner (1969), *J. Phys. Chem. Solids* **30**, 483.

69B3 H. Bartholin, D. Block, and F. Chaisse (1969), *C.R. Acad. Sci. B* **269**, 67.

69C1 G. R. Caskey, D. J. Sellmyer, and L. G. Rubin (1969), *Rev. Sci. Instrum.* **40**, 1280.

69C2 R. K. M. Chow and H. U. Everts (1969), *Phys. Rev.* **188**, 947.

69C3 M. C. Cadeville, F. Gautier, C. Robert, and J. Roussel (1969), *Solid State Commun.* **7**, 1701.

69F1 A. Fert (1969), *J. Phys. C* **2**, 1784.

69F2 K. Fischer (1969), *Z. Phys.* **225**, 444.

69H1 D. J. Huntley and C. W. E. Walker (1969), *Can. J. Phys.* **47**, 805.

69H2 A. J. Heeger (1969), *Solid State Physics,* Academic Press, Inc., New York, Vol. 23, 284.

69L1 J. O. Linde (1969), *Ark. Fys.* **39**, 139.

69M1 G. T. Meaden and N. H. Sze (1969), *J. Less Common Metals* **19**, 444.

69M2 K. Maki (1969), *Prog. Theor. Phys.* **41**, 586.

69P1 R. F. Powell (1969), *Br. J. Appl. Phys.* **2**, 1467.

69P2 A. B. Pippard (1969), *The Physics of Metals,* edited by J. M. Ziman, Cambridge University Press, Cambridge, 113.

69R1 E. R. Rumbo (1969), *Philos. Mag.* **19**, 689.

69S1 S. K. Srivastava and P. K. Sharma (1969), *Solid State Commun.* **7**, 601.

69S2 K. Schröder and A. Giannuzzi (1969), *Phys. Status Solidi* **34**, K133.

69S3 L. L. Sparks and W. J. Hall (1969), *U.S. Nat. Bur. Stand. Report* #9721.

69S4 D. Shoenberg (1969), *The Physics of Metals,* edited by J. M. Ziman, Cambridge University Press, Cambridge, 62.

69T1 H. J. Trodahl (1969), *Rev. Sci. Instrum.* **40**, 648.

69T2 H. J. Trodahl and F. J. Blatt (1969), *Phys. Rev.* **180**, 706.

69T3 P. V. Tamarin, G. E. Chuprikov, and S. S. Shalyt (1969), *Zh. Eksp. Teor. Fiz.* [*Sov. Phys.-JETP*] **28**, 836.

69V1 M. V. Vedernikov (1969), *Adv. Phys.* **18**, 337.

69V2 N. V. Volkenshtein, L. A. Ugodnikova and Yu. N. Tsiovkin (1969), *Zh. Eksp. Teor. Fiz. Pis'ma Red.* [*JETP Lett.*] **10**, 48.

69W1 G. Williams and J. W. Loram (1969), *J. Phys. Chem. Solids* **30**, 1827.

69W2 J. Woollam (1969), *Phys. Rev.* **185**, 995.

70A1 T. Aisaka and M. Shimizu (1970), *J. Phys. Soc. Japan* **28**, 646.

70A2 V. Allali, P. Donze, D. Gainon, and J. Sierro (1970), *J. Appl. Phys.* **41**, 1154.

70C1 R. Carter, A. Davidson, and P. A. Schroeder (1970), *J. Phys. Chem. Solids* **31**. 2374.

70C2 R. S. Crisp and J. Rungis (1970), *Philos. Mag.* **22**, 217.

70D1 A. R. DuCharme and L. R. Edwards (1970), *Phys. Rev. B* **2**, 2940.

70D2 J. Durand and F. Gautier (1970), *J. Phys. Chem. Solids* **31**, 2773.

70F1 R. H. Freeman and J. Bass (1970), *Rev. Sci. Instrum.* **41**, 1171.

70F2 T. Farrell and D. Greig (1970), *J. Phys. C* **3**, 138.

70F3 R. Fletcher, N. S. Ho, and F. D. Manchester (1970), *Metal Phys.* **3**, 59.

70F4 C. L. Foiles (1970), *Philos. Mag.* **21**, 1279.

70G1 I. C. Getting and G. C. Kennedy (1970), *J. Appl. Phys.* **41**, 4552.

70K1 J. F. Koch and R. F. Doezema (1970), *Phys. Rev. Lett.* **24**, 507.

70K2 A. B. Kaiser and S. Doniach (1970), *Int. J. Magn.* **1**, 11.

70L1 P. D. Long and R. E. Turner (1970), *J. Phys. C* **2**, S127.

70L2 J. W. Loram, T. G. Whall, and P. J. Ford (1970), *Phys. Rev. B* **2**, 857.

70M1 F. M. Mueller, A. J. Freeman, J. O. Dimmock, and A. M. Furdyna (1970), *Phys. Rev. B* **1**, 4617.

70N1 P. E. Nielsen and P. L. Taylor (1970), *Phys. Rev. Lett.* **25**, 371.

70N2 P. E. Nielsen, P. L. Taylor, and F. D. Manchester (1970), *Phys. Lett. A* **32**, 161.

70N3 V. F. Nemchenko, S. N. L'vov, P. I. Mal'ko, and N. P. Vereshchaka (1970), *Fiz. Met. Metalloved.* [*Phys. Met. Metallogr.*] **30**, 202.

70N4 I. Nagy and L. Pal (1970), *Phys. Rev. Lett.* **24**, 894.

70P1 R. L. Powell and D. H. Weitzel (1970), *J. Res. Nat. Bur. Stand. A* **74**, 673.

70R1 V. A. Rowe and P. A. Schroeder (1970), *J. Phys. Chem. Solids* **31**, 1.

70S1 F. C. Schwerer and L. J. Cuddy (1970), *Phys. Rev. B* **2**, 1575.

70S2 S. Skalski, M. P. Kawatra, J. A. Mydosh, and J. I. Budnick (1970), *Phys. Rev. B* **2**, 3613.

70T1 Philip L. Taylor (1970), *A Quantum Approach to the Solid State*, Prentice-Hall, Englewood Cliffs, New Jersey.

70W1 R. A. Weiner and M. T. Béal-Monod (1970), *Phys. Rev. B* **2**, 2675.

70Z1 J. E. Zimmerman, P. Thienne, and J. T. Harding (1970), *J. Appl. Phys.* **41**, 1572.

71A1 R. S. Averback and J. Bass (1971), *Phys. Rev. Lett.* **26**, 882.

71B1 R. D. Barnard (1971), *Phys. Status Solidi B* **46**, 369.

71B2 R. Berman and J. Kopp (1971), *J. Phys. F* **1**, 457.

71C1 L. Colquitt, H. R. Fankhauser, and F. J. Blatt (1971), *Phys. Rev. B* **4**, 292.

71C2 M. C. Cadeville and J. Roussel (1971), *J. Phys. F* **1**, 686.

71E1 G. J. Edwards (1971), *J. of Phys. E* **4**, 299.

71E2 J. W. Ekin and B. W. Maxfield (1971), *Phys. Rev. B* **4**, 4215.

71F1 A. Fert and I. A. Campbell (1971), *J. Phys. (Paris)* **32**, C1, 46.

71F2 K. Fischer (1971), *Phys. Status Solidi B* **46**, 11.

71F3 C. L. Foiles (1971), *J. Phys. Chem. Solids* **32**, 1205.

71G1 D. Gugan (1971), *Proc. Roy. Soc. A* **325**, 223.

71G2 A. M. Guénault (1971), *J. Phys. F* **1**, L1.

71H1 D. J. Huntley (1971), *Can. J. Phys.* **49**, 2610.

71K1 C. Kittel (1971), *Introduction to Solid State Physics*, 4th ed., John Wiley and Sons, Inc. New York.

71K2 I. Ya. Korenblit and Yu. P. Lazarenko (1971), *Zh. Eksp. Teor. Fiz.* [*Sov. Phys.-JETP*] **33**, 837.

71L1 D. Lazarus, R. N. Jeffery, and J. D. Weiss (1971), *Appl. Phys. Lett.* **19**, 371.

71M1 D. L. Mills, A. Fert, and I. A. Campbell (1971), *Phys. Rev. B* **4**, 196.

71M2 W. M. MacInnes and K. Schröder (1971), *Phys. Rev. B* **4**, 4091.

71N1 F. Napoli and D. Sherrington (1971), *J. Phys. F* **1**, L53.

71O1 Omega Engineering, Inc. (1971), *Thermocouple Calibration Tables and Alloy Data*, Omega Engineering, Stamford, Conn.

71P1 I. Pemberton and A. M. Guénault (1971), *Phys. Lett. A* **37**, 71.

71S1 S. C. Smith and A. C. Anderson (1971), *Cryogenics* **11**, 53.

71S2   H. Suhl (1971), *J. Phys. (Paris)* **32**, C1–421.

71S3   W. M. Star (1971), Thesis, Leiden.

71T1   S. H. Tang, P. P. Craig, and T. A. Kitchens (1971), *Phys. Rev. Lett.* **27**, 593.

71V1   N. V. Volkenshtein, V. A. Novoselov, and V. E. Startsev (1971), *Zh. Eksp. Teor. Fiz.*
       [*Sov. Phys.-JETP*] **33**, 584.

71V2   A. Von Middendorff (1971), *Cryogenics* **11**, 318.

71W1   B. Window (1971), *J. Phys. F* **1**, 533.

72A1   J. C. Allnut and A. J. Walton (1972), *J. Phys.* E **5**, 131.

72A2   B. E. Armstrong and R. Fletcher (1972), *Can. J. Phys.* **50**, 244.

72A3   R. S. Averback and D. K. Wagner (1972), *Solid State Commun.* **11**, 1109.

72B1   J. Bass (1972), *Adv. Phys.* **21**, 431.

72B2   R. D. Barnard (1972), *Thermoelectricity in Metals and Alloys,* Taylor and Francis Ltd.,
       London.

72B3   P. Blood and D. Greig (1972), *J. Phys. F* **2**, 79.

72B4   M. G. Brereton (1972), *Philos. Mag.* **25**, 1019.

72B5   F. J. Blatt (1972), *Can. J. Phys.* **50**, 2836.

72B6   F. F. Bekker (1972), *Phys. Lett. A* **41**, 301.

72B7   R. Berman (1972), *Temperature, Its Measurement and Control in Science and Industry,*
       Vol IV, 1537, Instrument Society of America, Pittsburgh, Pa.

72D1   A. W. Dudenhoeffer and R. R. Bourassa (1972), *Phys. Rev. B* **5**, 1651.

72G1   D. Greig, T. K. Brunk, and P. A. Schroeder (1972), *Philos. Mag.* **25**, 1009.

72G2   G. N. Granneman and L. Berger, (1972), *Bull. Am. Phys. Soc.* **17**, 235.

72G3   R. P. Giffard, R. A. Webb, and J. C. Wheatley (1972), *J. Low Temp. Phys.* **6**, 533.

72G4   A. M. Guénault (1972), *J. Phys. F* **2**, 316.

72H1   R. P. Huebener (1972), *Solid State Physics,* Vol 27, 63, Academic Press, New York.

72H2   J. G. Hust, R. L. Powell, and L. L. Sparks (1972), *Temperature, Its Measurement and
       Control in Science and Industry,* Vol IV, 1525, Instrument Society of America,
       Pittsburgh, Pa.

72L1   J. Langlinais and J. Callaway (1972), *Phys. Rev. B* **5**, 124.

72L2   C. C. Lee and P. A. Schroeder (1972), *Philos. Mag.* **25**, 1161.

72L3   W. E. Lawrence and J. W. Wilkins (1972), *Phys. Rev. B* **6**, 4466.

72N1   V. F. Nemchenko, S. N. L'vov, P. I. Mal'ko, and V. N. Deliyev (1972), *Physics of Metals
       and Metallog,* **33**, 82.

72S1   L. L. Sparks, R. L. Powell, and W. J. Hall (1972), *Reference Tables for Low Temperature
       Thermocouples,* Nat. Bur. Stand. Monograph 124.

72S2   L. L. Sparks and R. L. Powell (1972), *J. Res. Nat. Bur. Stand. A* **76**, 263.

72S3   R. L. Singh and G. T. Meaden (1972), *Phys. Status Solidi A* **9**, K173.

72S4   R. L. Singh and G. T. Meaden (1972), *Phys. Rev. B* **6**, 2660.

72T1   G. A. Thomas, K. Levin, and R. D. Parks (1972), *Phys. Rev. Lett.* **29**, 1321.

72W1   J. Waldman and M. B. Bever (1972), *Metall. Trans.* **3**, 1607.

73A1   R. S. Averback, C. H. Stephan, and J. Bass (1973), *J. Low Temp. Phys.* **12**, 319.

73A2   H. Auerbach, D. Flynn, S. Goetsh, C. C. Lee, and P. A. Schroeder (1973), *Philos. Mag.*
       28, 49.

73B1   R. R. Bourassa and A. W. Dudenhoeffer (1973), *Phys. Rev. B* **7**, 1270.

73F1   C. L. Foiles (1973), *Philos. Mag.* **27**, 757.

73G1   D. Greig and D. Livesey (1973), *J. Phys. F* **2**, 699.

73G2   J. C. Garland (1973), *Appl. Phys. Lett.* **22**, 203.

73H1   C. Haas (1973), *New Developments in Semiconductors,* edited by P. R. Wallace, R.
       Harris, and M. H. Zuckermann, Noordhoff Inst. Publ. Co., Leyden, 1.

73H2   S. Hufner, G. K. Wertheim, and J. H. Wernick (1973), *Phys. Rev. B* **8**, 4511.

73I1   *Temperature, Its Measurement and Control in Science and Industry,* Instrument Society of America (1973), Vol. IV.

73K1   P. A. Kinzie (1973), *Thermocouple Temperature Measurement,* John Wiley and Sons, New York.

73K2   T. Knittel (1973), *Cryogenics* **6**, 370.

73K3   P. G. Klemens (1973), *Physica* **69**, 171.

73M1   S. N. Mahajan, J. G. Daunt, R. I. Boughton, and M. Yaqub (1973), *J. Low Temp. Phys.* **12**, 347.

73P1   D. C. Price and G. Williams (1973), *J. Phys. F* **3**, 810.

73R1   R. B. Roberts (1973), Ph.D. Dissertation, University of Western Australia.

73R2   T. Rybka and R. R. Bourassa (1973), *Phys. Rev. B* **8**, 4449.

73R3   N. Rivier and K. Adkins (1973), in *Amorphous Magnetism,* edited by H. O. Hooper and A. M. deGraaf, Plenum Press, New York.

73S1   H. Suhl, ed., (1973), *Magnetism,* Vol V, Academic Press, New York.

73T1   H. J. Trodahl (1973), *J. Phys. F* **3**, 1972.

73Z1   I. Zoric, G. A. Thomas, and R. D. Parks (1973), *Phys. Rev. Lett.* **30**, 22.

74B1   E. Babic and J. R. Cooper, unpublished data.

74B2   F. J. Blatt, C. K. Chiang, and L. Smrčka (1974), *Phys. Status Solidi A* **24**, 621.

74B3   F. J. Blatt, A. D. Caplin, C. K. Chiang, and P. A. Schroeder (1974), *Solid State Commun.* **15**, 411.

74C1   R. S. Crisp, W. G. Henry, E. Whalley, and R. W. Wilson, unpublished data.

74C2   A. D. Caplin, C. Chiang, P. A. Schroeder, and J. Tracy (1974), *Phys. Status Solidi A* **26**, 497.

74C3   J. R. Cooper, unpublished data.

74C4   C. K. Chiang (1974), *Rev. Sci. Instrum.* **45**, 985.

74C5   A. D. Caplin, C. K. Chiang, and P. A. Schroeder (1974), *Phil. Mag.* **30**, 1177.

74F1   C. L. Foiles, unpublished results.

74F2   K. Fischer (1974), *J. Low Temp. Phys.* **17**, 87.

74G1   J. C. Garland (1974), *Proc. LT13,* Vol. 4, 399, Plenum Press, New York.

74G2   J. C. Garland and D. J. Van Harlingen (1974), *Phys. Rev. B* **10**, 4825.

74G3   V. F. Gantmakher (1974), *Rep. Prog. Phys.* **37**, 317.

74G4   D. Greig and J. A. Rowlands (1974), *J. Phys. F* **4**, 232.

74G5   A. M. Guénault (1974), *J. Phys. F* **4**, 256.

74G6   J. F. Goff, private communication.

74G7   A. M. Guénault (1974), *Philos. Mag.* **30**, 641.

74G8   F. Gautier, private communication.

74G9   J. E. Graebner, J. J. Rubin, R. J. Schutz, F. S. L. Hsu, W. A. Reed, and R. J. Higgins (1974), *Conference on Magnetism and Magnetic Materials,* Amer. Inst. Physics, New York, p. 445.

74H1   W. G. Henry, private communication.

74H2   A. Hasegawa (1974), *Solid State Commun.* **15**, 1361.

74H3   A. Hasegawa (1974), *J. Phys. F* **4**, 2164.

74K1   W. Kesternich and C. Papastaikouidis (1974), *Phys. Status Solidi B* **64**, K41.

74M1   K. Matho and M. T. Béal-Monod (1974), *J. Phys. F* **4**, 848.

74N1   P. E. Nielsen and P. L. Taylor (1974), *Phys. Rev. B* **10**, 4061.

74P1   C. Piotrowski, C. H. Stephan, and J. Bass (1974), *Proc. LT13,* Vol 4, 417, Plenum Press, New York.

74P2   R. L. Powell, W. J. Hall, C. H. Hyink, L. L. Sparks, G. W. Burns, M. G. Scroger, and H. H. Plumb (1974), *Thermocouple Reference Tables Based on the IPTS–68,* NBS. Monograph 125.

74P3   R. G. Poulsen, D. L. Randles, and M. Springford (1974), *J. Phys. F.* **4**, 981.

74R1   C. Rizzuto (1974), *Rep. Prog. Phys.* **37**, 147.

74S1   D. J. Sellmyer and J. M. Franz (1974), *Can. J. Phys.* **52**, 2060.

74S2   P. A. Schroeder, unpublished results.

74S3   H. H. Sample, L. J. Neuringer, and L. G. Rubin (1974), *Rev. Sci. Instrum.* **45**, 64.

74S4   D. J. Stanley (1974), *Proc. R. Soc. A* **339**, 97.

74S5   M. P. Sarachik and J. S. Tunger (1974), *Bull. Am. Phys. Soc.* **19**, 304.

74T1   J. Tracy and L. Smrčka, unpublished data.

74T2   J. Tracy, unpublished data.

74T3   S. H. Tang, T. H. Kitchens, F. J. Cadieu, and P. P. Craig (1974), *Proc. LT13*, Vol 4, 385, Plenum Press, New York.

74T4   J. Tracy (1974), *Phys. Lett. A*, **48**, 219.

74W1   Jonathan D. Weiss and David Lazarus (1974), *Phys. Rev. B* **10**, 456.

74Z1   V. Zlatić and N. Rivier (1974), *J. Phys. F* **4**, 732.

76H1   T. Holstein and S. K. Lyo (1976), private communication.

76O1   J. L. Opsal, B. J. Thaler, and J. Bass (1976), *Phys. Rev. Lett.* **16**, 1211.

| Author Index

251

Subject Index